Advanced Structured Materials

Volume 76

More information about this series at http://www.springer.com/series/8611

Mohammed A. Gondal · Chang Xiaofeng
Mohamed A. Dastageer

Novel Bismuth-Oxyhalide-Based Materials and Their Applications

Springer

Mohammed A. Gondal
Department of Physics
King Fahd University of Petroleum and Minerals
Dhahran
Saudi Arabia

Mohamed A. Dastageer
Department of Physics
King Fahd University of Petroleum and Minerals
Dhahran
Saudi Arabia

Chang Xiaofeng
Department of Applied Chemistry
Nanjing University of Aeronautics and Astronautics
Nanjing
China

ISSN 1869-8433 ISSN 1869-8441 (electronic)
Advanced Structured Materials
ISBN 978-81-322-3737-2 ISBN 978-81-322-3739-6 (eBook)
DOI 10.1007/978-81-322-3739-6

Library of Congress Control Number: 2017940815

Printed on acid-free paper

This Springer imprint is published by Springer Nature
The registered company is Springer (India) Pvt. Ltd.
The registered company address is: 7th Floor, Vijaya Building, 17 Barakhamba Road, New Delhi 110 001, India

Preface

Energy and water are the two commodities that the modern world is running short of due to over population, over consumption, and vast industrialization and all the major world confrontations were and will be for controlling and manipulating these precious natural resources. Water is the foundation of life and more than 70% of planet earth is covered with water, but quite ironically, nearly 1.2 billion people on earth do not have access to clean drinking water. The rapid industrialization and reckless human activities wreak havoc on the precious water bodies by dumping alarming amount of organic, inorganic, and biological pollutants into it. Over the past few years, a variety of strategies have been investigated to develop efficient and pollution free technologies for the demineralization of highly toxic water pollutants to nontoxic products for waste water treatment. Photocatalysis has been emerged as one of the most promising technologies as it embodies more economical and green way to demineralize pollutants by using the energy from sunlight or artificial light sources. Photocatalytic process has a great potential for the removal of organic pollutants (dyes, phenolic compounds, pesticides, and herbicides), heavy metals, and other toxic pathogens (bacteria, fungus candida) from water. In addition photocatalysis can be applied for solar energy harvesting based on photovoltaic principle, especially in dye sensitized solar cells.

Ever since, Fujishima and Honda reported water splitting using TiO_2 electrode, many pure, doped, and composite variants of TiO_2 have been synthesized and used for various photocatalytic and solar energy applications due to its many positive attributes like low cost, physical and chemical stability, and high efficiency. However, one of the major constraints of TiO_2 for any photocatalytic application is its wide band gap energy, which limits its activity in the visible spectral region and hence it cannot be a good photocatalyst under solar radiation. Many pure, doped, and composite catalysts of CdS, SnO_2, WO_3, SiO_2, ZnO, Nb_2O_3, and Fe_2O_3 have been synthesized and used for various applications under solar radiation. In recent days a new breed of photocatalysts, bismuth oxyhalides have attracted considerable interest due to their efficient photocatalytic activity, high stability, low cost, and other properties like magnetic separation after use in water. These efforts have been focused on those bismuth oxyhalide (BiOX, X= Cl, Br, I) based photocatalysts which are highly efficient, cheap, and could absorb in the visible region.

In this book, we have focused on recent development on various methods of synthesis of BIOX, their morphological and optical characterizations, different variants of doped and composites of BiOCl, BiOBr, BiOI, and their use in the photocatalytic waste water treatment, secondary batteries, gas sensors, and solar cells. Also this book touches upon the future prospect of these materials, in terms of synthesis, multidimensional shape dependence, and the hetero junctions of these materials for various photocatalytic and other applications. BiOX is considered to be very promising photocatalyst due to its utility of visible light as an excitation source and thus the use of solar radiations for future field applications for waste water treatment and solar energy harvesting.

The authors would like to thank the Deanship of Scientific Research, King Fahd University of Petroleum and Minerals in supporting this book project under project # IN141024. The authors would like to thank Mr Praveen Kumar from Springer for his patience and encouragement and Mr. Tanweer Alam from KFUPM who has worked on the formatting of the book. The authors are indebted to many people who have contributed to the improvement of this book by pointing to the errors and possible revisions. We thank many authors and publishers who were gracious enough to grant permission to use the figures from their works. The international Conference on Water, Environment, Energy and Society authors will appreciate any future cooperation of readers for improving this book.

King Fahd University of Petroleum and Minerals
Dhahran, Saudi Arabia

Mohammed A. Gondal
Chang Xiaofeng
Mohamed A. Dastageer

Acknowledgements

The authors would like to acknowledge the financial support provided by the Deanship of Scientific Research at King Fahd University of Petroleum and Minerals (KFUPM) for funding this book writing grant through project number IN141024. The authors are also grateful for the support received from the Department of Physics at KFUPM and in particular all the members and students working in the Laser Research Group of Physics Department.

The authors would like to thank Mr. Praveen Kumar from Springer Nature for his patience and encouragement. The authors would also like to thank Mr. Tanweer Alam from KFUPM for his support in the formatting of the book. We thank many authors and publishers who were gracious enough to grant permission to use figures and illustrations from their works. The authors are indebted to many people who have contributed to the improvement of this book by pointing to the errors and possible revisions.

Contents

About the Authors

Dr. Mohammed A. Gondal is a distinguished professor at Physics Department, King Fahd University of Petroleum and Minerals, Dhahran Saudi Arabia and visiting researcher at MIT, USA since 2012. Dr. Gondal took his Ph.D. degree from the University of Bonn, Germany. He has been a regular associate member of the International Center for Theoretical Physics, Trieste, Italy (1988–1995) and member of the editorial boards of seven international journals. Dr. Gondal's research interests are lasers, photonics, nanotechnology, materials science, sensors, solar cells, laser remote sensing, pollution monitoring, development of laser-based analytical techniques (LIDAR, PAS, LIBS, LEI), methane cracking using photolysis, laser-photocatalysis for production of high value hydrocarbons and hydrogen for fuel cells, waste water treatment, and disinfection of bacteria. Dr. Gondal has published over 450 research papers in ISI journals and conferences of high repute. He has 15 patents so far published by US patent office. He has contributed numerous book chapters. Dr. Gondal was awarded Al-Marai Innovation Prize for the year 2011 for the application of nano-structured materials for environmental, three times Distinguished (Best) researcher award, best project award by KFUPM.

Mr. Chang Xiaofeng is a Ph.D. research scholar at College of Materials Science and Technology, Nanjing University of Aeronautics and Astronautics, Nanjing, China. Mr. Chang has published over 50 research papers in ISI journals and conference proceedings. He has numerous patents and good citations.

Mr. Mohamed A. Dastageer is currently a lecturer in Physics and scientist at Laser Research Group at King Fahd University of Petroleum and Minerals (KFUPM), Saudi Arabia. He took his master degree in Physics from King Fahd University of Petroleum and Minerals, Saudi Arabia. His areas of specialization are Laser Molecular Spectroscopy, Optical Properties of Functional Materials and Photocatalysis for the production of high value hydrocarbons, waste water treatment, and disinfection of bacteria. He is a research team member of KFUPM-MIT, USA and KFUPM-NUS, Singapore collaborative research projects. He has several research publications in reputed international journals, book chapters, and a few patents to his credit.

Chapter 1
Introduction to Bismuth Oxyhalides

Abstract This chapter is the general introduction to bismuth oxyhalides as a functional material, with the particular emphasis for their application as a photocatalyst in many photochemical processes. This introductory chapter touches upon the importance of bismuth oxyhalides, their unique characteristics, and different variants of these materials for various applications.

Keywords Functional materials · Bismuth oxyhalides · Photocatalysis · Layered structure

1.1 Background

The relation between man and material is as old as the history of known human civilizations, which is evident from the fact that the historical ages are known by the materials used by the people of that age. In the recent evolution in the application of materials, man transformed from being satisfied and content with whatever is naturally available, to embarking on synthesizing the material of his specific requirements by modifying the very atomic and molecular characteristics of the materials. The material revolution brought about by the development of nanoscience and technology expanded the horizon of the material science that churns out innumerable breeds of new materials with remarkable characteristics, which had been unknown to our predecessors. The modern way of carrying out applied research starts with the sequence of synthesizing application oriented materials, studying their morphological, mechanical, thermal, electrical, magnetic, and optical characteristics with state-of-the-art analytical tools and optimizing the material characteristics to get the best out of this material for the particular application. For example, ever since, the photoelectrochemical water splitting using rutile TiO_2 single crystals were reported in 1972 by Fujishima and Honda [1]; this method and material were extended to various photocatalytic applications in the field of energy research and environmental remediation. Although TiO_2 is highly photoactive, the optical band-gap energy of this material lies in the UV spectral

M.A. Gondal et al., *Novel Bismuth-Oxyhalide-Based Materials and Their Applications*, Advanced Structured Materials 76,
DOI 10.1007/978-81-322-3739-6_1

region of the solar radiation, which not only limits the efficiency of the material but also the effective harvesting of solar energy. Hence, wide scientific and material research was directed in the pursuit of visible light active materials, which are capable of working with the visible spectral region of solar radiation, in addition to promoting other remarkable photocatalytic characteristics [2]. In order to transform a semiconducting material visible light active, the band structure of the material needs to be engineered to achieve reduced band-gap energy and for this methods like doping, junction formation, alloying, coupling with solar absorbing materials of lower band-gap energies, and developing new breed of functional materials are adopted [3–7].

1.2 Importance of Bismuth-Based Functional Materials

Bismuth-based materials are less toxic, very abundant, and were proven to be very good photocatalytic material in the visible spectral region, particularly in the application of water purification [8–11]. Bismuth-based materials become visible active due to the formation of favorable hybridized valence band between Bi 6s in Bi(III) and O 2p levels. Bismuth Oxyhalides, BiOX (X = Cl, Br, I), are the new kind of layered materials exhibiting excellent optical and electrical characteristic and are capable for various applications including photocatalysis. BiOX are a class of V–VI–VII ternary semiconductors, having matlockite structure, in which [X–Bi–O–Bi–X] layers are stacked and held by Vader Walls interaction through the halogen atom, whereas all the atoms within the layer are covalently bonded and this structure leads up to excellent electrical optical and mechanical properties. In each [X–Bi–O–Bi–X] layer, Bi atom is surrounded by four each of oxygen and halogen atoms resulting in asymmetric decahedral symmetry. Many variants are BiOX nanostructures such as nanosheets, nanobelts, nanofibers, nanowires, and nano-flowers have been synthesized using various chemical routes [12–14]. In recent days, BiOX materials have been used for many photocatalytic degradation studies as these materials possess the inherent tendency for low electron hole recombination, which emerge from the open crystalline structure, indirect band-gap energy, and the existence of an electric field directing perpendicular to the slab, that helps in the separation of photogenerated electron hole pairs.

Although, BiOX materials are proven to be the excellent photocatalysts, their photcatalytic behaviors highly depend on their structural characteristics like size, shape, dimensions, and crystal phases. Hence, achieving such BiOX material with remarkable photocatalytic attributes depends on the different methods of syntheses which are adopted to get the well-defined nanostructures such as one-dimensional, two-dimensional, three-dimensional, hierarchical, and hallow nanostructures that result in suitable band structure to absorb radiation to initiate the photocatalytic reaction [15]. Recent works on the synthesis of BiOX showed that the liquid-phase synthesis methods like hydrothermal, solvothermal, templated, sonochemical, interface-mediated, reverse phase microemulsion-based, and anion exchange-based synthesis are very effective in producing the BiOX nanomaterials with desired functionality. Among these many synthesis methods, hydrothermal and solvothermal methods are most widely used, where the characteristics of BiOX can be controlled by adjusting the thermodynamic and kinetic parameters of the synthetic system. Also BiOX nanomaterials of any desired nanostructures can be realized by adopting template synthesis route, where the nanostructured templates of desired structural features are used.

In the case of liquid-phase synthesis route of BiOX nanostructures, both the bismuth and halogen sources are dispersed in aqueous solution, where the Bi^{3+} cations react with water molecule to form $(Bi_2O_2)^{2+}$ and this further combines with X^- anions through Coulomb interaction to form nuclei of X–Bi–O–Bi–X slabs. These tiny slab of X–Bi–O–Bi–X are stacked together by Van der waals interaction after prolong ripening time [16, 17]. The kinetics of above self-assembly is vastly influenced by reaction environment such as solvents, additives, impurities, reaction temperature, and the reaction time. Also in order to achieve the architecture of BiOX nanostructures, the fast chemical reaction between $(Bi_2O_2)^{2+}$ and X^- anions should be inhibited to reduce the spontaneous formation of two-dimensional building blocks. The rate of this fast reaction can be significantly reduced by adopting methods by adding external agents like acids in the bismuth-containing aqueous solution or by obtaining bismuth-containing coordination compounds or using ethylene glycol as the reaction solvent [18]. Also the acidity in the growth environment plays a vital role in deciding the crystal facet exposure and phase transformation. For example, the growth of {001} facet of BiOCl can be inhibited and favor the generation of {010} facet by controlling the pH the reaction solution and it was reported that {001} facet of BiOCl is UV active whereas {010} facet of BiOCl is visible active in the process of photocatalytic degradation of methyl orange dye [19]. Also by precisely tuning the pH value of the growth environment of BiOX nanocrystals, we can transform them into new layered structure that could result in a favorable band structure that lead up to improved light absorbance and the charge-carrier transfer.

1.3 Unique Characteristics of Bismuth Oxyhalides

BiOX nanocrystals have been on the research spot light due to their favorable band structures capable of absorbing sunlight to carryout photocatalytic reactions. However, the inherent limitations of these materials like their inability to work efficiently in the visible region of the solar radiation and the increased tendency for the recombination of photogenerated charge-carriers. In order to overcome these limitations and make these materials an effective photocatalyst, various strategies have been adopted. One method is by manipulating the internal electric field induced by the unique layered structure of BiOX in order to facilitate the separation and transfer of photogenerated charge-carriers and thereby enhancing the photocatalytic activity of the material [20]. It was reported that the higher exposure of

{001} facets significantly increased the internal electric field and helped the charge separation to enhance phtocatalytic activity [19]. The second strategy to improve the functionality of BiOX is by removing the lattice halogen atom by replacing it with OH^- or by heating it at very high temperature or by solid-state displacement reaction to form a non-stoichiometric layered structure and this change in the structure results in corresponding change in the band structure and consequently the optical absorption edge and the band redox potential change [19]. The third method to alter the electronic properties of BiOX semiconductor to create productive light harvesting and reduced charge combination is by creating the oxygen vacancies on the surface of the material [21] by using the functional organic molecule on the surface of the BiOX [22], which serve as the electron or hole scavenger in order to reduce the electron hole recombination.

The most commonly adopted approach to modify the absorption and the charge separation characteristics of any semiconducting material is doping with some foreign atom, but in the case of BiOX semiconductors such doping was found to be deteriorating the functionality of the material [23]. However, doping with other halogen atom in BiOX or self doping with the same halogen atom to certain extent improved the performance of the material [24]. Also anchoring of nanostructured noble metals like silver on the BiOX semiconductor was tried to induce surface plasmon resonance (SPR) effect in the material and it was found that the coupling of such metallic nanoparticles on BiOX increased the optical absorption and the separation and transfer of photogenerated charge-carriers [25]. This SPR effect takes place due to the resonant effect of the electric field created by the oscillation of free electron with the electric fields of the exciting electromagnetic radiation and this effect is so delicate and depends on the shape, size and the arrangement of metal nanoparticles. Also the broad range of band redox potential of BiOX ensures that these materials can easily be a good partner with wide variety of semiconducting material to form the nanocomposite [26]. These composite partners form a heterojunction that enables the photogenerated electrons and holes to spatially separate, and thereby reduce the charge recombination probability. BiOX can also form a heterojunction with carbon-based material like graphene, where graphene can function as a co-catalyst to extract photogenerated electrons and this will inhibit the electron hole recombination to increase the photocatalylic efficiency of BiOX [27].

The unique layered structure of BiOX renders the scope of this material for various applications such as environmental remediation and energy conversion. However, most of reports on the photocatalytic application of BiOX have been restricted to the degradation of organic pollutants and its applications in the area of water splitting and CO_2 reduction into value added hydrocarbons have not yet been focused. It should be noted that unlike the versatile and the traditional catalysts like TiO_2 and WO_3, BiOX is still in the early stage of research in the field of photocatalysis and has a good scope for improving it as a very functional photocatalysts and also for various other potential applications. The interlacing $[Bi_2O_2]$ slabs with double halogen slabs coupled with a strong ionic nature of the material can promote exchange reactions between BiOX and the incoming atomic species and this promises that this material can be used for wider applications. The theoretical models studying the influence of internal electric field in the layered structure of BiOX on the photogenerated charge transfer, the effect of different facets of BiOX on the intensity of this internal electric field have a great scope for the research and development of this functional material.

This book describes various synthetic routes of BiOX, their unique chemical structure, band structure, optical characteristics, morphological characteristics, and various modified variants of BiOX. In the application side of BiOX, the book touches upon their applications on photocatalytic hydrogen production, environmental remediation, water purification, solar cells, and fuel cells. The book concludes with the views on future direction and the prospects of this functional material.

References

1. Fujishima K Honda (1972) Electrochemical photolysis of water at a semiconductor electrode. Nature 238:37–38
2. Kudo Akihiko, Miseki Yugo (2009) Heterogeneous photocatalyst materials for water splitting. Chem Soc Rev 38:253–278
3. Banerjee S, Pillai SC, Falaras P, O'Shea KE, Byrne JA, Dionysiou DD (2014) New insights into the mechanism of visible light photocatalysis. J Phys Chem Lett 5:2543–2554
4. Kim S, Hwang S-J, Choi W (2005) Visible light active platinum-ion-doped TiO_2 photocatalyst. J Phys Chem B 109:24260–24267
5. Etacheri V, Seery MK, Hinder SJ, Pillai SC (2010) Highly visible light active TiO_2–x N x Heterojunction photocatalysts†. Chem Mater 22:3843–3853
6. Bhachu DS, Sathasivam S, Carmalt CJ, Parkin IP (2014) PbO-Modified TiO_2 thin films: a route to visible light photocatalysts. Langmuir 30:624–630
7. Asahi R, Morikawa T, Ohwaki T, Aoki K, Taga Y (2001) Visible-light photocatalysis in nitrogen-doped titanium oxides. Science 293:269–271
8. Zhou L, Wang WZ, Xu HL, Sun SM, Shang M (2009) The influence of g-C_3N_4 loading on the photocatalytic activity of $Bi_{12}O_{17}Br_2/Bi_2O_3$ composite in the phenol red degradation. Chem Eur J 15:1776–1782
9. Cheng HF, Huang BB, Lu JB, Wang ZY, Xu B, Qin XY, Zhang XY, Dai Y (2010) Synergistic effect of crystal and electronic structures on the visible-light-driven photocatalytic performances of Bi_2O_3 polymorphs. Phys Chem Chem Phys 12:15468–15475
10. Yao WF, Xu XH, Wang H, Zhou JT, Yang XN, Zhang Y, Shang SX, Huang BB (2004) Photocatalytic property of perovskite bismuth titanate. Appl Catal B 52:109–116
11. Cheng HF, Huang BB, Yang KS, Wang ZY, Qin XY, Zhang XY, Dai Y (2010) Facile template-free synthesis of $Bi_2O_2CO_3$ hierarchical microflowers and their associated photocatalytic activity. Chem Phys Chem 11:2167–2173
12. Wang CH, Shao CL, Liu YC, Zhang LN (2008) Photocatalytic properties BiOCl and Bi_2O_3 nanofibers prepared by electrospinning. Scr Mater 59:332–335
13. Ye LQ, Zan L, Tian L, Peng T, Zhang J (2011) The 001 facets-dependent high photoactivity of BiOCl nanosheets. Chem Commun 47:6951–6953
14. Zhang D, Li J, Wang QG, Wu QS (2013) High 001 facets dominated BiOBr lamellas: facile hydrolysis preparation and selective visible-light photocatalytic activity†. J Mater Chem A 1:8622–8629
15. Osterloh FE (2013) Inorganic nanostructures for photoelectrochemical and photocatalytic water splitting. Chem Soc Rev 42:2294
16. Yin YD, Rioux RM, Erdonmez CK, Hughes S, Somorjai GA, Alivisatos AP (2004) Formation of hollow nanocrystals through the nanoscale Kirkendall effect. Science 304:711

17. Fan HJ, Knez M, Scholz R, Nielsch K, Pippel E, Hesse D, Zacharias M, Gösele U (2006) Monocrystalline spinel nanotube fabrication based on the Kirkendall effect. Nat Mater 5:627
18. Zhang X, Ai ZH, Jia FL, Zhang LZ (2008) Generalized one-pot synthesis, characterization, and photocatalytic activity of hierarchical BiOX (X = Cl, Br, I) nanoplate microspheres. J Phys Chem C 112:747
19. Zhao K, Zhang LZ, Wang JJ, Li QX, He WW, Yin JJ (2013) Surface structure-dependent molecular oxygen activation of BiOCl single-crystalline nanosheets. J Am Chem Soc 135:15750
20. Jiang J, Zhao K, Xiao XY, Zhang LZ (2012) Synthesis and facet-dependent photoreactivity of BiOCl single-crystalline nanosheets. J Am Chem Soc 134:4473
21. Wahlström E, Vestergaard EK, Schaub R, Rønnau A, Vestergaard M, Lægsgaard E, Stensgaard I, Besenbacher F (2004) Electron transfer-induced dynamics of oxygen molecules on the TiO2(110) surface. Science 303:511
22. Wang YN, Deng KJ, Zhang LZ (2011) Visible light photocatalysis of BiOI and its photocatalytic activity enhancement by in situ ionic liquid modification. J Phys Chem C 115:14300
23. Yu JH, Wei B, Zhu L, Gao H, Sun WJ, Xu LL (2013) Flowerlike C-doped BiOCl nanostructures: facile wet chemical fabrication and enhanced UV photocatalytic properties. Appl Surf Sci 284:497
24. Zhang X, Zhang LZ (2010) Electronic and band structure tuning of ternary semiconductor photocatalysts by self doping: the case of BiOI. J Phys Chem C 114:18198
25. Linic S, Christopher P, Ingram DB (2011) Plasmonic-metal nanostructures for efficient conversion of solar to chemical energy. Nat Mater 10:911
26. Smith AM, Nie SM (2010) Semiconductor nanocrystals: structure, properties, and band gap engineering. Acc Chem Res 43:190
27. Xiang QJ, Yu JG, Jaroniec M (2012) Graphene-based semiconductor photocatalysts. Chem Soc Rev 41:782

Chapter 2
Microstructure, Morphology, and Surface Structure of Bismuth Oxyhalides

Abstract This chapter presents the tetragonal matlockite structure of bismuth oxyhalides and the results of X-ray diffraction, Raman spectra, and FTIR spectra of the material. The surface chemical analysis using XPS results, surface morphology using high-resolution TEM images of bismuth oxyhalides are also presented.

Keywords Matlockite structure · XRD · XPS · Raman spectra · FTIR · HRTEM

2.1 Space Group and Atomic Coordination (XRD Analysis)

All the BiOX molecules exhibit tetragonal matlockite structure, which characteristically crystallize to PbFX—type tetragonal structure with P4/nmm space group. In the unit cell of BiOX, each Bi atom is coordinated by four oxygen atoms and four halogen atoms. In this structure each oxygen atom is attached to four neighboring bismuth atoms, forming a tetragonal pyramid with the oxygen atom at its apex and also each halogen atom is linked with the neighboring bismuth atoms to form tetragonal pyramid with halogen atom at its apex. Also these molecules are of a layered structure characterized by $[Bi_2O_2]$ slabs interleaved by double slabs of halogen atoms as shown in Fig. 2.1. The layers of BiOX are attached by the Van der Waals interaction through the X atoms along the *c*-axis. These unique layered structures provide a large space for the polarization of the atoms and orbitals, thereby the electron hole pairs can be efficiently separated by the induced dipole. The crystal model along the [001], [010], and [110] directions are shown in Figs. 2.2, 2.3, and 2.4, respectively. The atomic positions of BiOX crystals are summarized in Table 2.1.

As an example, let us consider the structure of BiOCl molecule illustrated in Figs. 2.1b and 2.4b, where the layered structure characterized by $[Bi_2O_2]$ slabs is interleaved by double slabs of chlorine atoms [1]. The stacking of [BiOCl] layers is through Van der waals interactions between the Cl atoms along the [001] direction [2]. Due to the characteristics of the layered structure, discussed above, the 2-D

M.A. Gondal et al., *Novel Bismuth-Oxyhalide-Based Materials and Their Applications*, Advanced Structured Materials 76,
DOI 10.1007/978-81-322-3739-6_2

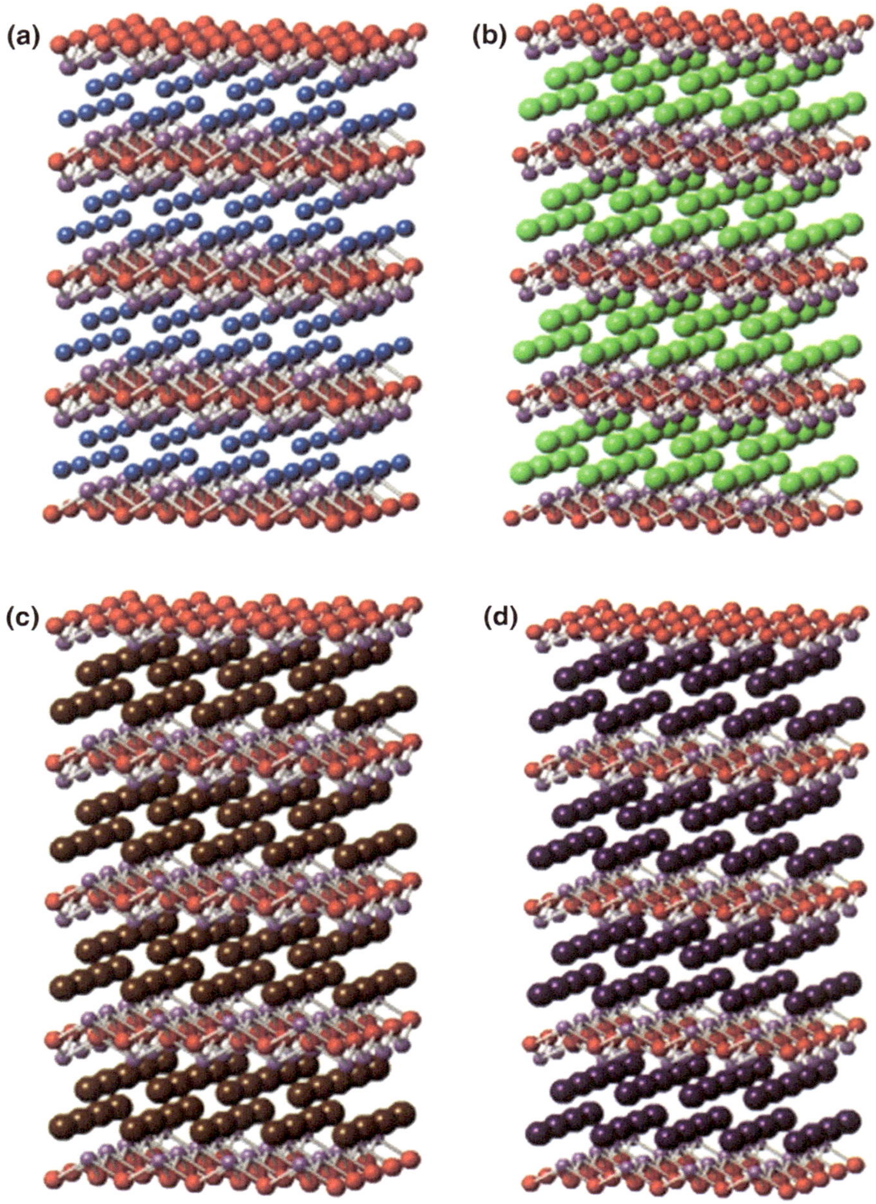

Fig. 2.1 The three-dimensional structure model illustration of BiOX crystals

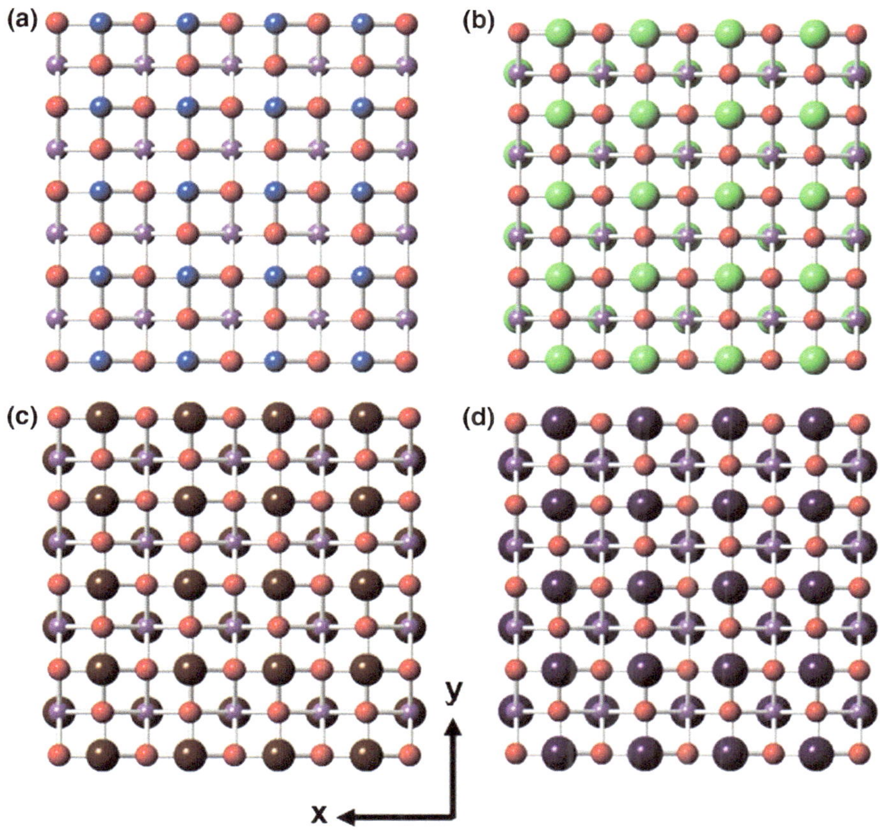

Fig. 2.2 The crystal model illustration along the [001] direction of BiOX crystals

laminar BiOCl nanoplates can easily be formed. Also in the case of BiOCl, the oxygen density (OD) is very high on the (001) and (110) facets as illustrated in Figs. 2.2b and 2.4b, which can be calculated as shown below: For 001 lattice and the top face, the oxygen density $OD_{(001)} = (1 + 4 \times ¼)/a^2 = 2a^{-2}$. Also BiOCl prepared by Wang et al. [3] showed a tetragonal structure with lattice constants $a = b = 3.883$ Å and $c = 7.347$ Å, where c is 1.892 times of a. From these data, the oxygen density $OD_{(110)}$ on the (110) facets can be calculated to be $0.7474a^{-2}$ ($OD_{(110)} = (2 * ½ + 4 * ¼)/(2^{1/2} * a * c) = 2/(2.676a^2) = 0.7474a^{-2}$) and the oxygen density on (001) facet is 2.68 times that on the (110) facets, which clearly indicates that the oxygen density on the (001) facets is much higher than that on the (110) facets.

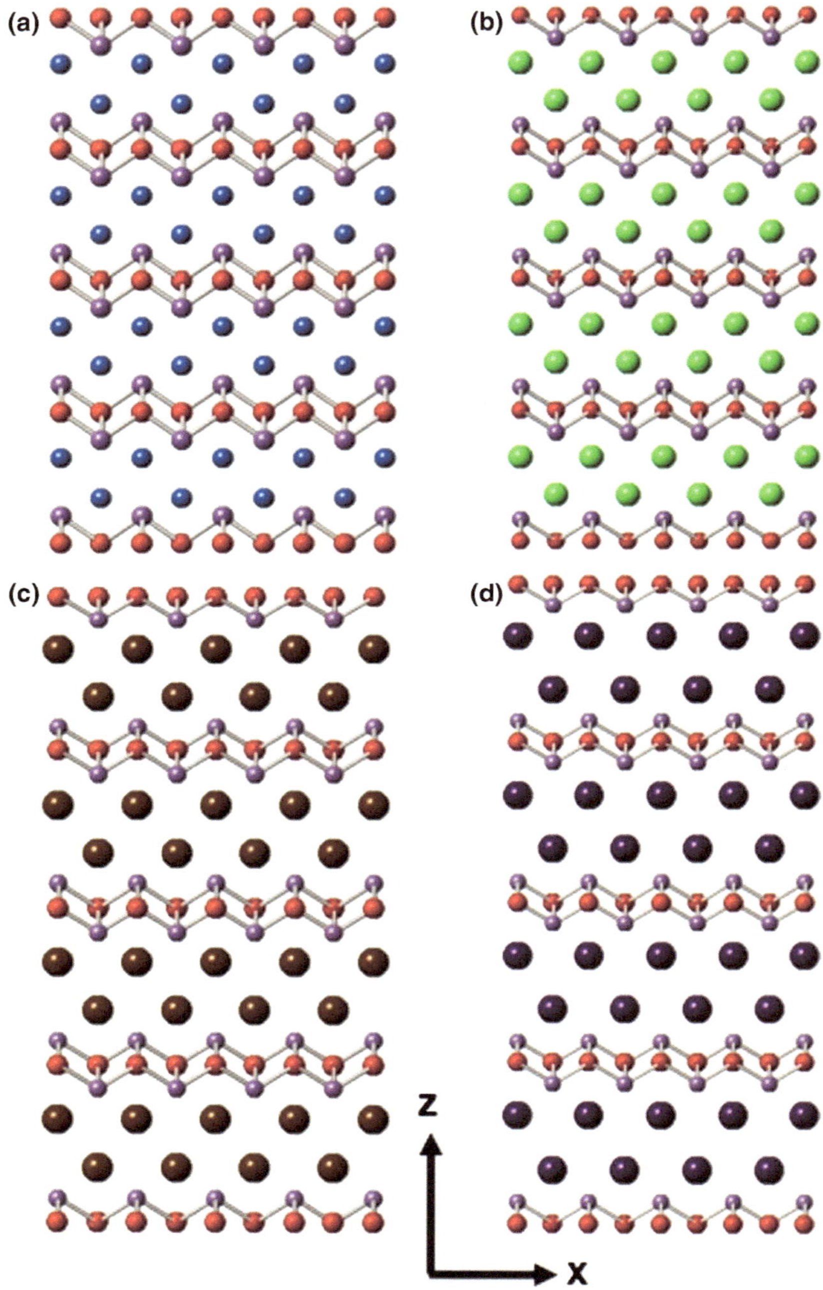

Fig. 2.3 The crystal model illustration along the [010] direction of BiOX crystals

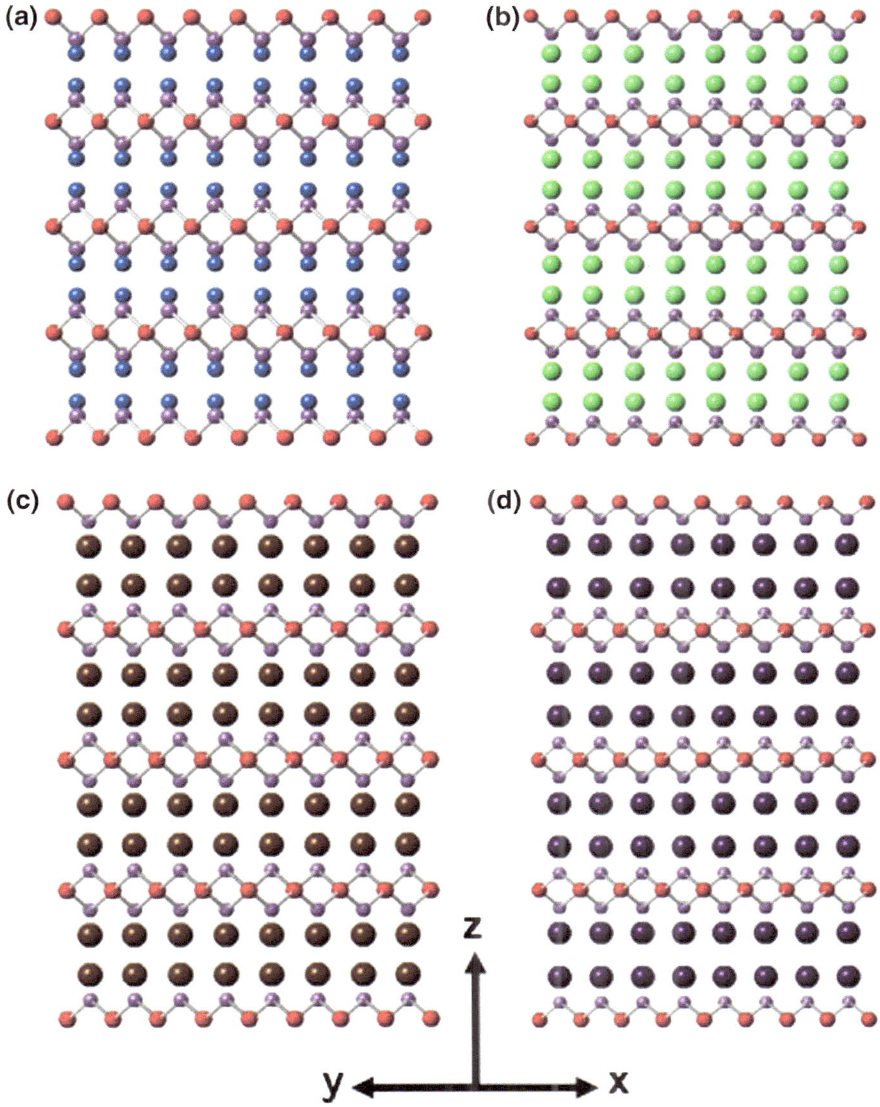

Fig. 2.4 The crystal model illustration along the [110] direction of BiOX crystals

The angles (φ) between different (*hkl*) planes in the tetragonal BiOCl are calculated using the experimentally determined lattice constants (a, b, c) and the Miller indices (h, k, l) in the following equation:

Table 2.1 Atomic positions of BiOX crystals

Atomic coordination	BiOF			BiOCl			BiOBr			BiOI		
	Bi	O	F	Bi	O	Cl	Bi	O	Br	Bi	O	I
x	0	0	0	0	0	0	0	0	0	0	0	0
y	0.5	0	0.5	0.5	0	0.5	0	0.5	0	0.5	0	0.5
z	0.208	0	0.65	0.17	0	0.645	0.154	0	0.653	0.132	0	0.668

$$\cos\varphi = \frac{\frac{1}{a^2}(h_1h_2 + k_1k_2) + \frac{1}{c^2}l_1l_2}{\sqrt{\left[\frac{1}{a^2}\left(h_1^2 + k_1^2\right) + \frac{1}{c^2}l_1^2\right]\left[\frac{1}{a^2}\left(h_2^2 + k_2^2\right) + \frac{1}{c^2}l_2^2\right]}}.$$

Also the mean particle size of BiOX samples along a certain crystallographic direction [*hkl*] can be calculated using the full width half maximum (*B*) of the XRD peak, the angle of diffraction (θ), and the wavelength of X-ray radiation (λ = 0.15406 nm) in the Scherrer equation: $D_{hkl} = K\lambda/(B\cos\theta)$, where *K* is constant in the equation whose value is 0.9.

Jiang et al. [4] synthesized BiOCl using $Bi(NO_3)_3{\cdot}5H_2O$ as a bismuth source and KCl as a halogen source and adjusting the pH value of solution to 6.0 by adding NaOH solution. From the crystallographic data of the synthesized materials, the theoretical values for the angles between (110) and (200) planes and between (002) and (102) planes were found to be 45° and 43.4°, respectively.

2.2 Raman and FTIR Analysis

It is clear from the structure that each primitive cell of BiOX has 6 monoatomic sites, which accounts for the 15 modes of vibration from the structure of reduced representation. Out of these 15 modes of vibrations, only 3 vibrational species (A_{1g}, B_{1g} and E_{1g}) are Raman active and only 2 vibrational species (A_{2u} and E_{1u}) are infrared active.

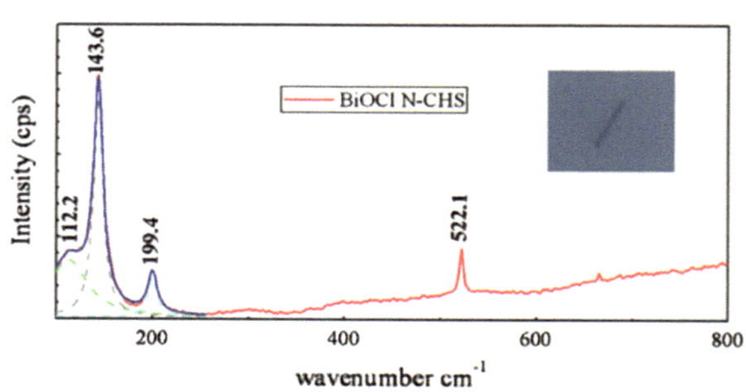

Fig. 2.5 Raman spectra of BiOCl. Reprinted with permission from Ref. [5]

Tian et al. [5] synthesized BiOCl nanowire by combining wet etch and liquid-phase crystal growth process and the Raman spectra of their material showed both Raman active modes A_{1g} (internal Bi–Cl stretching) at 143.6 cm^{-1} and E_{1g} (internal Bi–Cl stretching mode) at 199.4 cm^{-1} in Fig. 2.5.

Also Raman spectra of the BiOCl synthesized by Wang et al. [6] by annealing $BiCl_3$ and their material recorded Raman lines at 143.5 cm^{-1} (A_{1g} internal Bi–Cl stretching), 199.7 cm^{-1} (E_{1g} internal Bi–Cl stretching mode) and 394.8 cm^{-1} (B_{1g} due to motion of the oxygen atoms) in Fig. 2.6.

Almost the same Raman result was obtained by Xiong et al. [7] with their BiOCl synthesized by hydrothermal method. Raman lines at 145, 201, and 399 cm^{-1} were assigned to A_{1g} and E_{1g} and B_{1g}, respectively, as depicted in Fig. 2.7.

In the case of FTIR, although two vibrational species (A_{2u} and E_{1u}) are expected to be infrared active, only valent symmetrical A_{2u}-type vibrations of Bi–O at 505 cm^{-1} have been reported so for. In the work of Wang et al. [8], BiOCl was prepared by calcining polymer/inorganic composite fibers, fabricated by electrospinning method. Their FTIR results the only peak pertained to BiOCl observed was at 515 cm^{-1} due to A_{2u}-type valent symmetrical vibrations of Bi–O bond as indicated in Fig. 2.8.

Wang et al. [9] prepared the ionic liquid-modified BiOI (IL-BiOI) by reacting bismuth nitrate with [Bmim]I, where FTIR peak due to due to A_{2u}-type valent symmetrical vibrations of Bi–O bond was observed at 505 cm^{-1}. In addition to the peak of BiOCl, other FTIR peaks observed were at 1630 and 3450 cm^{-1}, assigned, respectively, to the bending and stretching vibrations of the hydroxyl group. Other FTIR peaks shown in Fig. 2.9 originate from the different modes of vibrations of organic compounds; for example, the peaks of 3146 cm^{-1} is due to the C–C bond in imidazolium ring, the two peaks at 2859 and 2926 cm^{-1} are from the symmetric and asymmetric stretch of the H–C–H bond in butyl group, the peak at 1380 cm^{-1} is due to the stretching modes of CN heterocycles and the peaks at 1170, 815, and 714 cm^{-1} are from the plane deformation and out of plane deformation of C–H bonds.

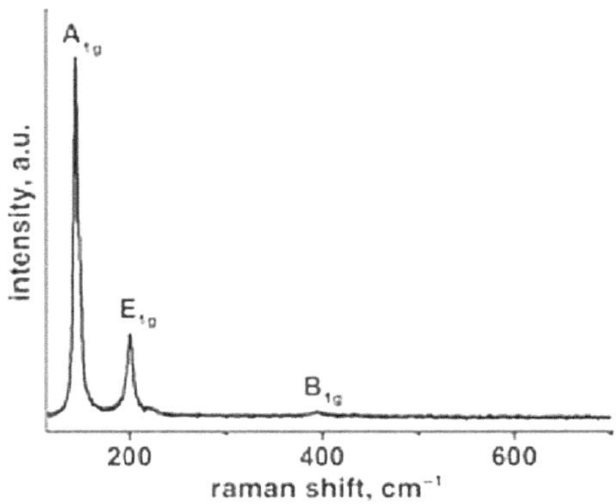

Fig. 2.6 Raman spectra of BiOCl. Reprinted with permission from Ref. [6]

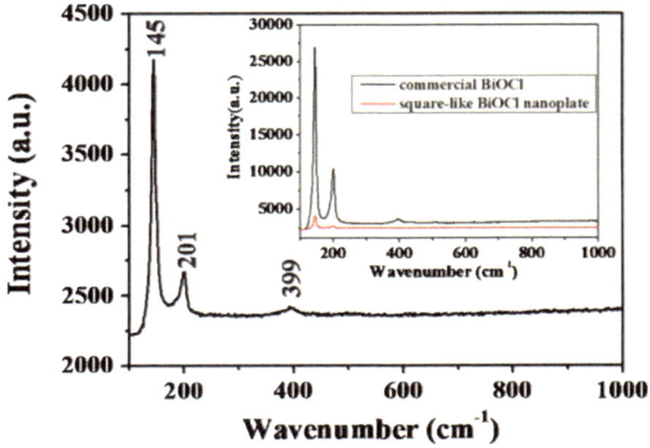

Fig. 2.7 Raman spectra of the as-synthesized BiO. Reprinted with permission from Ref. [7]

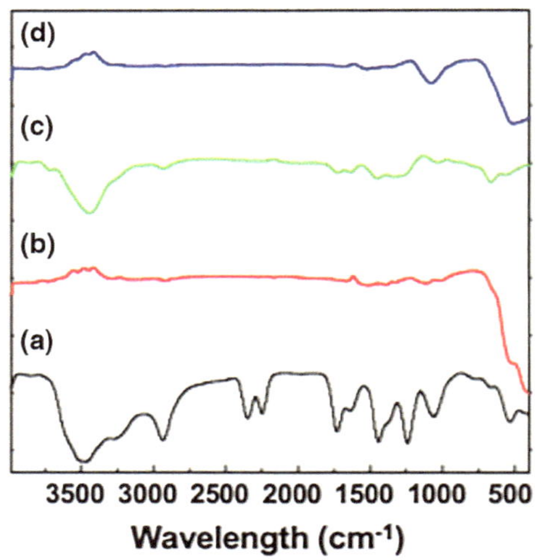

Fig. 2.8 FTIR spectra of **a** as-spun PAN/BiCl3 nanofibers, **b** BiOCl nanofibers, **c** as-spun PAN/Bi$(NO_3)_3$ nanofibers, **d** Bi_2O_3 nanofibers. Reprinted with permission from Ref. [8]

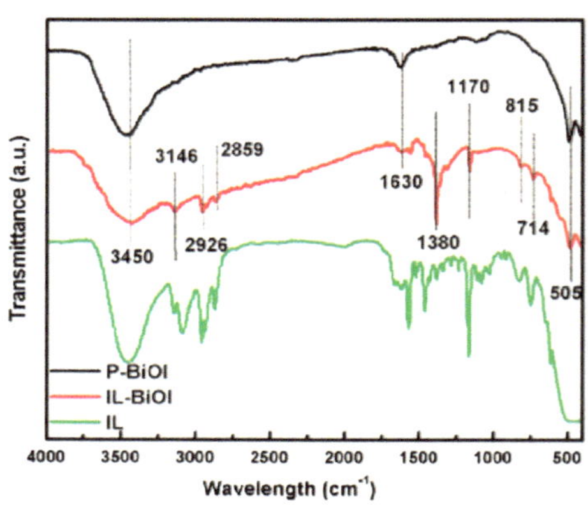

Fig. 2.9 FTIR spectra of P-BiOI, IL-BiOI and ionic liquid. Reprinted with permission from Ref. [9]

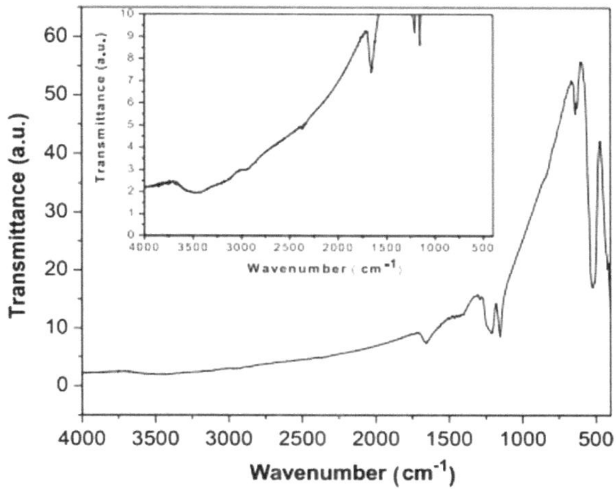

Fig. 2.10 FTIR spectrum of the BiOCl nanoplates. Reprinted with permission from Ref. [10]

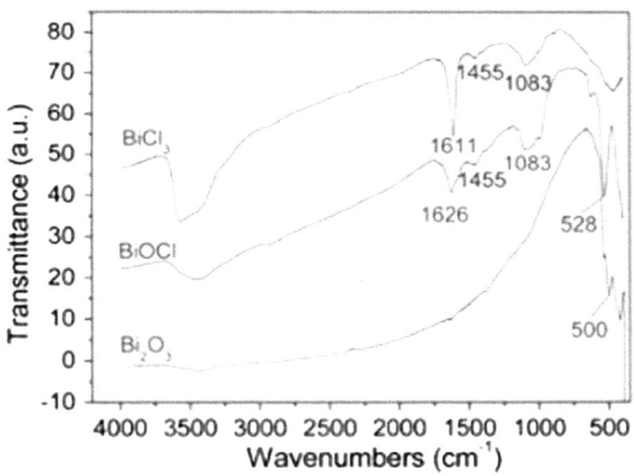

Fig. 2.11 FTIR spectrum of the BiOCl nanoplates. Reprinted with permission from Ref. [11]

Deng et al. [10] synthesized BiOCl by oxidizing bismuth particle hydrogen in the presence of sodium chloride and in their FTIR results, in Fig. 2.10, the peak due to A_{2u}-type valent symmetrical vibrations of Bi–O bond was observed at 505 cm^{-1} and other peaks around 3500 and 1650 cm^{-1} are due to the stretching and bending vibrations of the hydroxyl group, respectively.

Song et al. [11] synthesized flower-like BiOCl through solvothermal using pyridine as the solvent. Their FTIR results were a bit different from the one reported in other works, where peaks due to asymmetry and symmetric stretching vibration of Bi–Cl were observed, respectively, at 1455 and 1038 cm^{-1} in addition to the peak due to A_{2u}-type valent symmetrical vibrations of Bi–O bond at 505 cm^{-1}.

Also peaks around 3445 and 1626 cm^{-1} were assigned to, respectively, the stretching and bending vibrations of the hydroxyl group (Fig. 2.11).

2.3 Surface Chemical Structure

X-ray photoelectron spectroscopy (XPS), also known as electron spectroscopy for chemical analysis (ESCA), is a spectroscopic technique for elemental analysis by exciting both valence electrons and inner core electrons in an atom using the X-ray and also mapping their binding energies. In the case of BiOX, in addition to Bi, O, and X peaks, the peaks due to other elements originating from the precursors and other reactants used in a particular method of synthesis. As described in the earlier section, Wang et al. [8] first synthesized polymer/inorganic composite fibers by electrospinning method and subsequently BiOCl were prepared by the calcinations of the fiber. The XPS study of their end product naturally exhibited the peaks of Bi, O, Cl, and C, where carbon peak originated from the intermediate product. As shown in Fig. 2.12, the XPS peaks with binding energies 159.7 and 164.9 eV are, respectively, assigned to the Bi $4f_{7/2}$ and Bi $4f_{5/2}$ of BiOCl and the deconvoluted peaks at 158.6 and 163.8 eV are attributed also to Bi $4f_{7/2}$ and Bi $4f_{5/2}$, but from bismuth oxide, formed during the calcination of PAN/$BiCl_3$ precursor. Also the two peaks at 198 and 200.2 eV are, respectively, for the Cl $2p_{3/2}$ and Cl $2p_{1/2}$ of BiOCl.

Armelao et al. [12] prepared BiOCl by the controlled hydrolysis of $BiCl_3$ in the presence of acetylacetone, and their XPS scan revealed the presence of Bi, O, Cl, and C peaks. The peaks with binding energies 530.1 and 532 eV are assigned to the O 1s and the peaks at 159.1 and 198.1 eV correspond to Bi $4f_{7/2}$ and Cl 2p regions for BiOCl, respectively, as shown in Fig. 2.13.

BiOI SCN-4 was synthesized by Ye et al. [13] by annealing BiI_3 and their XPS study as shown in Fig. 2.14 indicating the presence of Bi, O, I, and C-related peaks. The Bi peaks for BiOI SCN-4, Bi $4f_{7/2}$, and Bi $4f_{5/2}$ are, respectively, at the binding

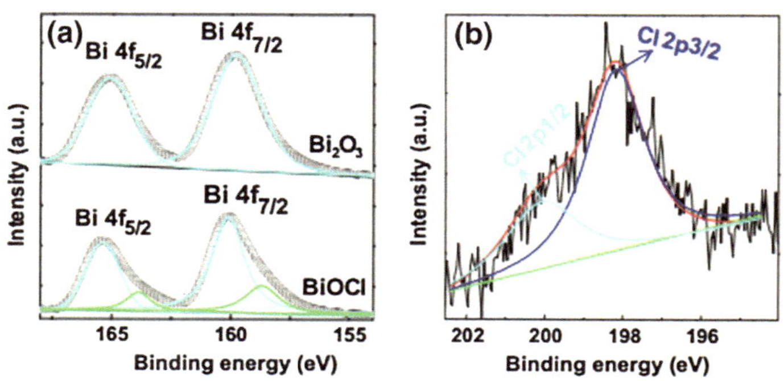

Fig. 2.12 High-resolution XPS survey spectra of Bi 4f in BiOCl and Bi_2O_3 nanofibers (**a**); XPS survey spectra of Cl 2p in BiOCl nanofibers (**b**). Reprinted with permission from Ref. [8]

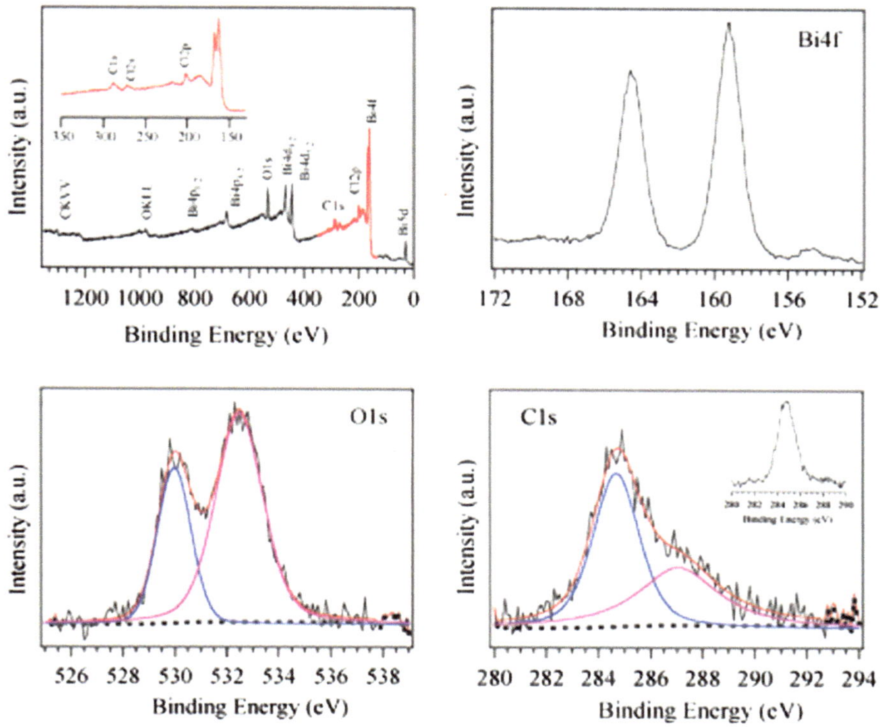

Fig. 2.13 High-resolution XPS spectra of as-prepared BiOCl. Reprinted with permission from Ref. [12]

energies of 159.1 and 164.4 eV and the peaks at 618.9 and 630.4 eV are attributed to I $3d_{5/2}$ and I $3d_{3/2}$ region, respectively. Also the peaks with binding energies of 530.1 and 532 eV are assigned to the O 1s for BiOI SCN-4.

Wang et al. [9] prepared the ionic liquid-modified BiOI (IL-BiOI), from the reaction of bismuth nitrate and [Bmim]I. The XPS analysis in Fig. 2.15 shows the presence of carbon peak in addition to the expected Bi-, O-, and I-related peaks. The two peaks due to C 1s are at the 284.6 and 286.2 eV binding energies, the two iodine peaks shown at 617.7 and 629.2 eV are attributed to I $3d_{5/2}$ and I $3d_{3/2}$ region of BiOI (IL-BiOI). The peaks with binding energy of 158.4 and 163.7 eV are assigned to the Bi $4f_{7/2}$ and Bi $4f_{5/2}$ and the peaks at 529.7 and 531.3 eV are assigned to the O 1s for BiOI (IL-BiOI).

Song et al. [11] adopted the solvothermal route to prepare flower-like BiOCl and their XPS survey on BiOCl exhibited the presence of Bi, O, Cl, and C peaks. In Fig. 2.16, in addition to the usual Bi $4f_{7/2}$ and Bi $4f_{5/2}$ peaks of BiOCl, three carbon peaks C 1s Cl $2p_{3/2}$ and Cl $2p_{1/2}$ region were also observed at 284.5, 197.7, and 199.2 eV, respectively. Also peaks with binding energy of 530.2 eV is assigned to the O 1s for BiOCl.

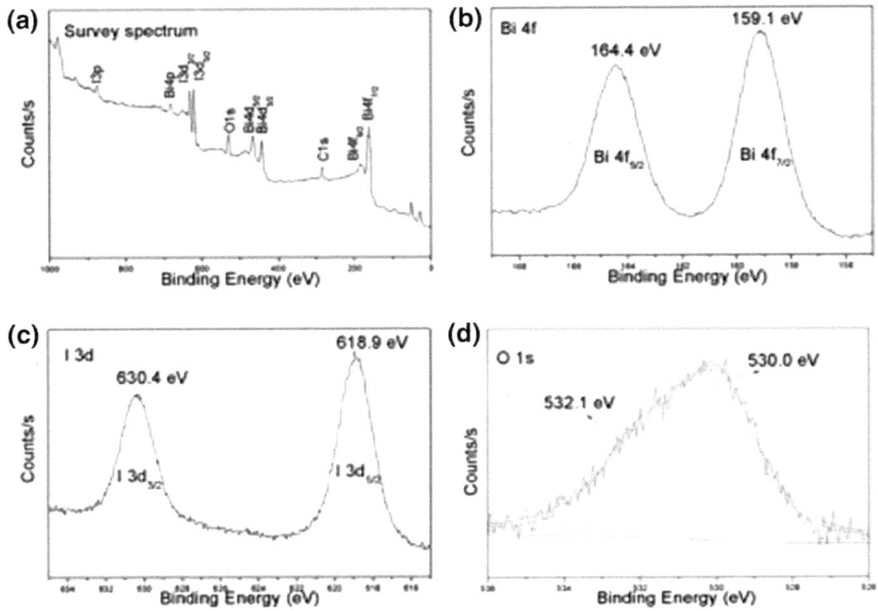

Fig. 2.14 Survey XPS spectrum (**a**) and high-resolution XPS spectra of BiOI SCN-4: Bi 4f (**b**), I 3d (**c**) and O 1s (**d**). Reprinted with permission from Ref. [13]

Xia et al. [14] synthesized BiOBr through a solvothermal process by employing an EG solution containing ionic liquid [C16mim]Br as the solvent, and as expected XPS survey scan of the BiOBr exhibited the presence of Bi-, O-, Br-, and C-related peaks. Besides Bi $4f_{7/2}$, Bi $4f_{5/2}$, and O 1s peaks for BiOBr, two peaks of Br 3d for BiOBr are also observed at around 67.95 eV in Fig. 2.17.

2.4 HRTEM Images and Analysis

Over the past decades, state-of-the-art electron microscopy has revolutionized our understanding of materials by revealing the crystal and electronic structural details down to subatomic and milli-electron-volt resolution [15–17]. In this regard, the work presented here begins with the evaluation of enhanced photodegradation onto a model composite photocatalyst of Bi/BiOCl synthesized by photoreduction method. Subsequently, the optical absorption caused by SPR effect and direct band gap of single bismuth nanoparticle that grew in situ on BiOCl nanosheet under the excitation of inelastic electron beam (e-beam) in a transmission electron microscope (TEM) have been well distinguished using state-of-the-art monochromated electron energy loss spectroscopy in scanning TEM imaging mode (Mono-STEM-EELS) with near atomic spatial resolution, which is also called spatially resolved valence

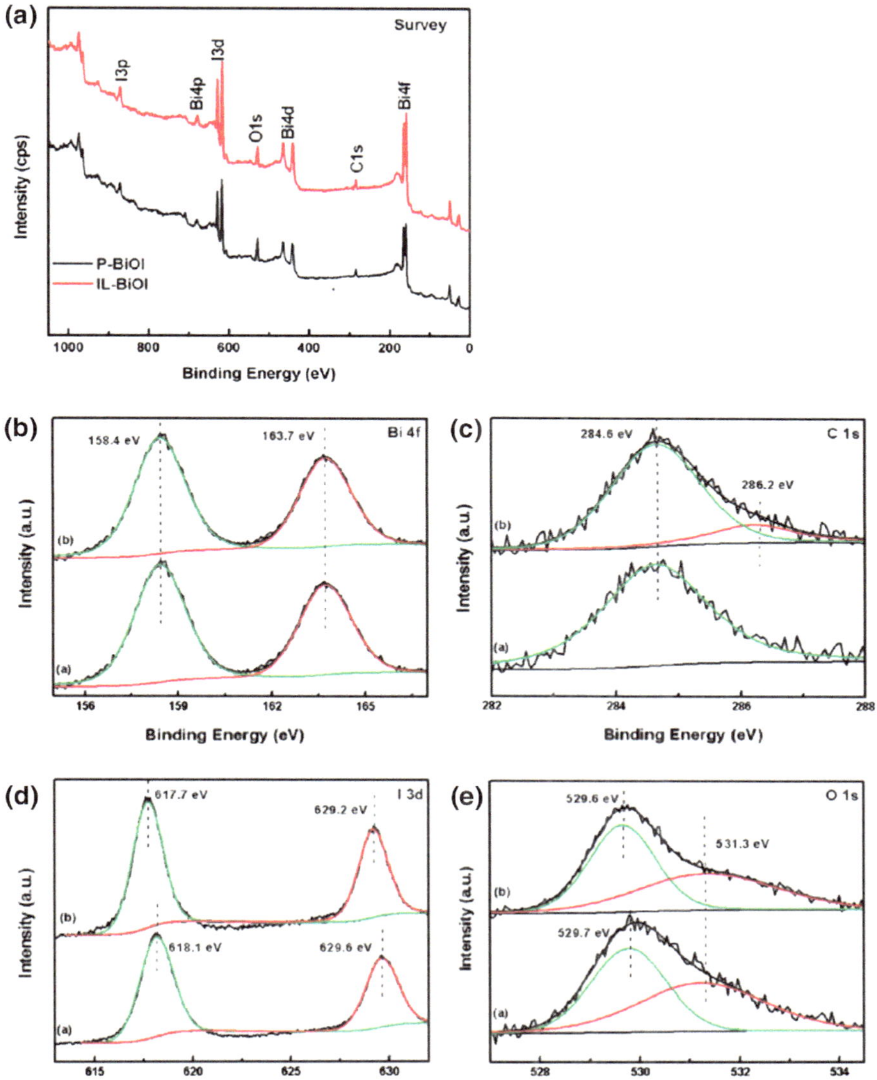

Fig. 2.15 XPS survey spectrum of the as-prepared BiOI samples (**a**). The high-resolution XPS spectra of (*a*) P-BiOI and (*b*) IL-BiOI samples in the regions of Bi 4f (**b**), C 1s (**c**), I 3d (**d**), and O 1s (**e**). Reprinted with permission from Ref. [9]

electron energy loss spectroscopy (VEELS) technique [18]. Furthermore, the charge transfer at the interface of Bi NPs and BiOCl has also been reasonably substantiated as well based on the band structural parameters obtained by Mono-STEM-EELS measurement. Hence, the possible mechanism behind improved photoreactivity over Bi/BiOCl catalyst was probed for the first time using the advanced electron

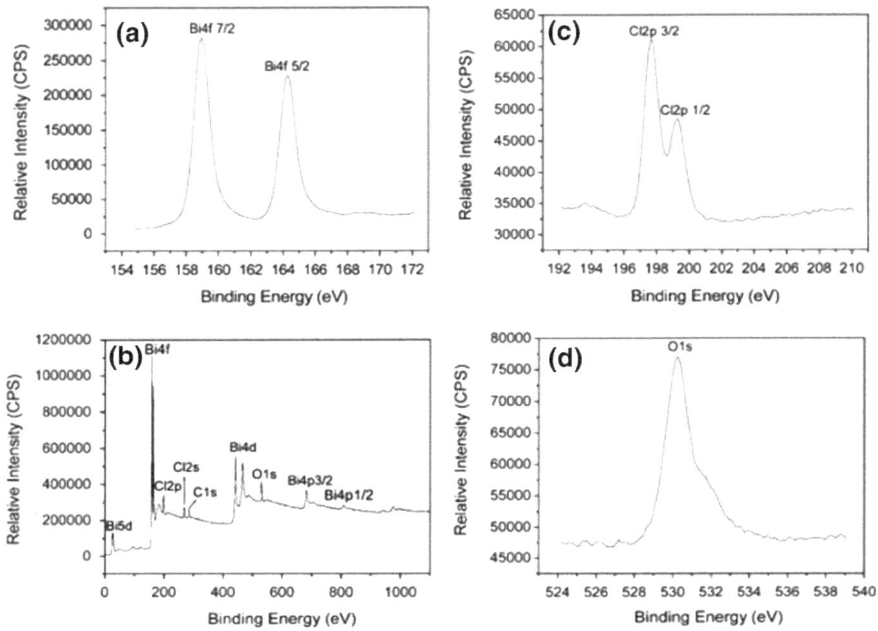

Fig. 2.16 XPS spectra of the BiOCl **a** XPS survey pattern of the sample, **b** XPS spectrum of the Bi 4f region, **c** XPS spectrum of the Cl 2p region, **d** XPS spectrum of the O 1s region. Reprinted with permission from Ref. [11]

microscopy techniques [19]. In our work [19], the direct synthesis of Bi/BiOCl in TEM was conducted before addressing the mechanism behind the improved photoreactivity of Bi/BiOCl catalyst by electron microscopy analysis. Stimulated by an adequate energy transfer from the electrons in an electron microscope under high vacuum, as shown in Fig. 2.18a–e and in situ TEM video (supplementary data), the nucleation, growth, and coalescence of Bi NPs onto BiOCl can be observed in real time [19]. The surface of BiOCl was clean and smooth initially, however, with a prolonged e-beam irradiation, the nucleation of a few Bi NPs can be found throughout the entire irradiated region, followed by a multiple coalescence among nanoparticles. The particle morphology continuously changed until Bi NPs with 2-D-projected hexagonal shape were observed. In addition, the atomic step has also been observed on the (110) surface of bismuth nanoparticle during the e-beam-induced growing process (Fig. 2.18f), which has been considered to play a significant role in the enhanced photocatalytic performance [20]. Unfortunately, an ex-situ TEM investigation, as shown in Fig. 2.2, shows that the surface reconstruction of the as-grown Bi NPs cannot be avoided after exposing to air and room temperature. The oxidation kinetics of Bi NPs is thermodynamically spontaneous from 200 to 1600 K and strongly dependent on the temperature [21]; but nevertheless, no apparent surface oxidation is observed from the HRTEM images of individual bismuth nanoparticle, indicating the relative chemical stability and potential application as a heterogeneous Bi NPs hybridized photocatalyst. As depicted in

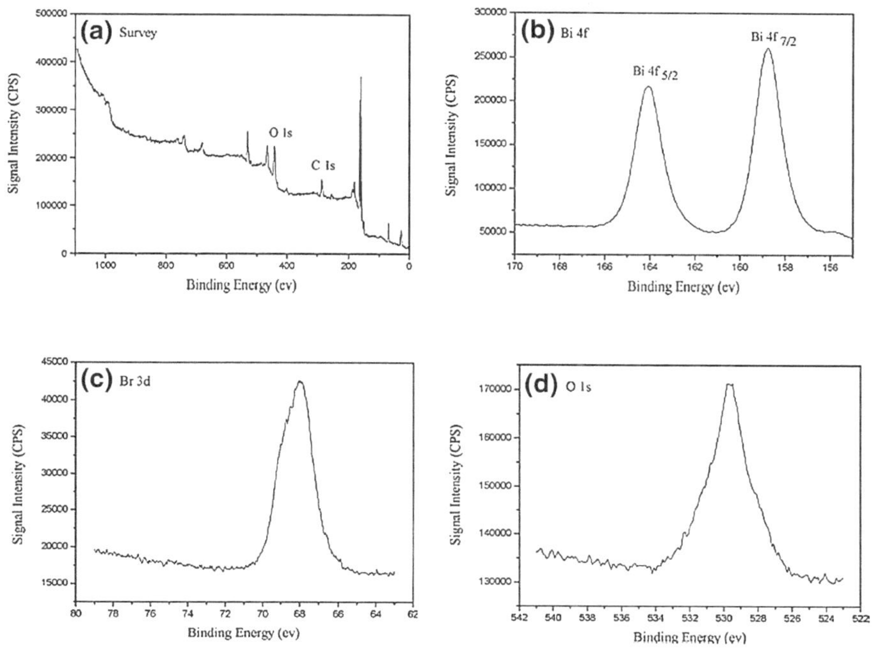

Fig. 2.17 XPS spectra of BiOBr **a** survey of the sample, **b** Bi 4f, **c** Br 3d, **d** O 1s. Reprinted with permission from Ref. [14]

Fig. 2.20, the low-loss Mono-STEM-EELS spectra [19] along with Gaussian fittings, extracted from the surface-loss EELS of a single bismuth nanoparticle, clearly show the energy loss peak at 5.5 eV (equivalent to 225 nm). The longer e-beam exposure time led to the further growth of Bi NPs, and the extinction cross section was in proportion to the third power of particle size according to the Mie theory [22]. Thereby, surface energy loss at 5.5 eV was noticeable in EELS signal, and was progressively intensified as the e-beam irradiations increased to 40 min.

Low-loss Mono-STEM-EELS maps were obtained by scanning probe in *x*–*y* direction across the region of interests, and the locations with strong EELS response (corresponds to the high brightness in the map) indicate their strongly active optical modes. As shown in Fig. 2.21, the Mono-STEM-EELS maps with different energies of Bi/BiOCl catalyst prepared under e-beam irradiation for 40 min further approve that the energy losses for surface plasmons are largely spread over 5–7 eV. Additionally, the surface-loss EELS extracted at 5.5 eV was also found to gradually expand from the specimen's surface to the vacuum as the growth of Bi NPs took place. The optical extinction spectra have been simulated and further compared with the surface-loss EELS results. It is reasonable to believe that the as-prepared single bismuth nanoparticle may have the hexagonal prism 3D structure in the light of HRTEM images (as depicted in Figs. 2.19e and 2.20) and

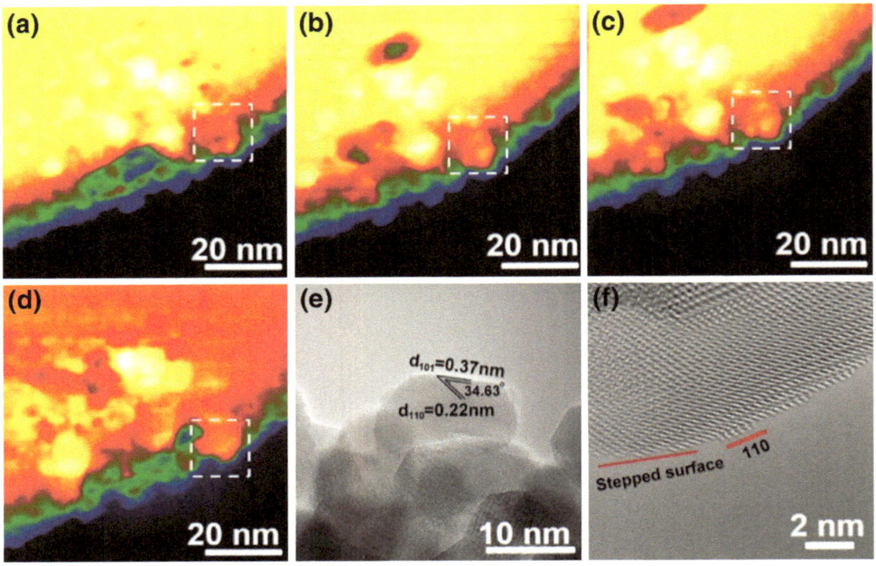

Fig. 2.18 High-angle annular dark field imaging in scanning TEM (STEM-HAADF) images (in false color) of single-crystalline BiOCl exposed by e-beam (average dose rate: ca. 0.5 e/A2 s) for 5 min (**a**), 10 min (**b**), 20 min (**c**), and 40 min (**d**), respectively. The *dashed squares* in (**a–d**) indicate the growth and coalescence of Bi NPs. **e** Depicts the HRTEM image of an individual bismuth nanoparticle and **f** shows the aberration-corrected HRTEM image near a (110) surface plane with a stepped surface (reference chang eta Applied catalysis B). Reprinted with permission from Ref. [19]

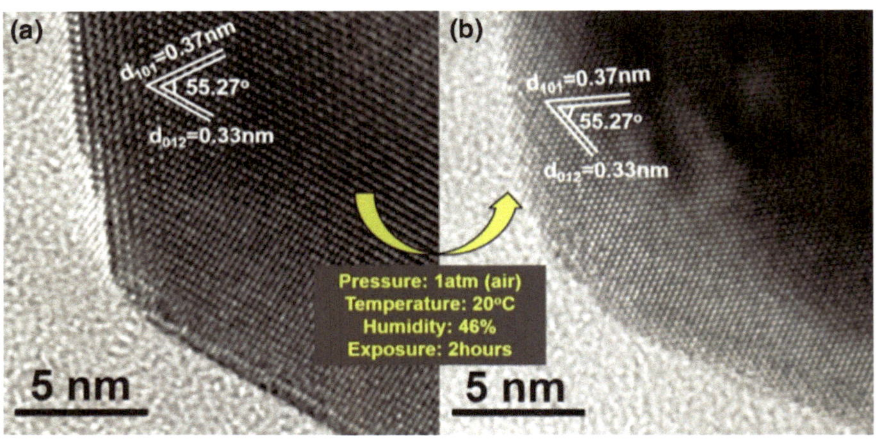

Fig. 2.19 Ex-situ TEM characterization: HRTEM images of the fresh grown bismuth nanoparticle induced by e-beam irradiation before (**a**) and after (**b**) exposing in air (humidity: 46%) at 20 °C for 2 h. The bismuth nanoparticle was slightly tilted possibly owning to the vibration during the extraction and reloading of sample holder in ex-situ experiment. Reprinted with permission from Ref. [19]

HAADF images taken with different α-tilt degrees. For the purposes of simplification, the simulation was conducted assuming a sphere with diameter of 20 nm, or a hexagonal prism with edge length and height of 10 and 20 nm, respectively. For a single bismuth nanoparticle with a spherical or hexagonal morphology, as depicted in Fig. 2.22, a pronounced excitation peak at ca. 260 nm can be observed, which suggests a local surface plasmon resonance and agrees with the peak in the Mono-STEM-EELS results. The electric field distribution around 5.5 eV suggests the surface plasmonic absorption peak is mainly dominated by the fundamental dipolar mode. The peak shifting between the optical extinction and experimental Mono-STEM-EELS spectra can be understood by the following two aspects: (1) the refractive index adopted in the simulation may deviate from that in the real sample [23]; (2) The e-beam can be regarded as a dipole source that is different from plane waves used in extinction spectra calculation. Under a dipole source (e-beam) incidence, the high-order mode of bismuth scatters [24]. The low-loss Mono-STEM-EELS signals (i.e., from the bulk of a single bismuth nanoparticle), as shown in Fig. 2.6 suggested a much smaller energy loss at around 1.4 eV that was not detected noticeably in the surface-loss EELS (Fig. 2.21, supplementary data, Ref. [20]). The E_g in bismuth nanoparticle could be assigned to the significant quantum confinement [25]. The energy loss near the band edge for a crystalline semiconductor in EELS spectra can be described by the following equation,

$$I = I_0 + K(E - E_g)n \quad (1)$$

where I and I_0 are the relative intensity (measured in counts) in EELS and the corresponding baseline, respectively. E, E_g, K, and n are the low energy loss (eV),

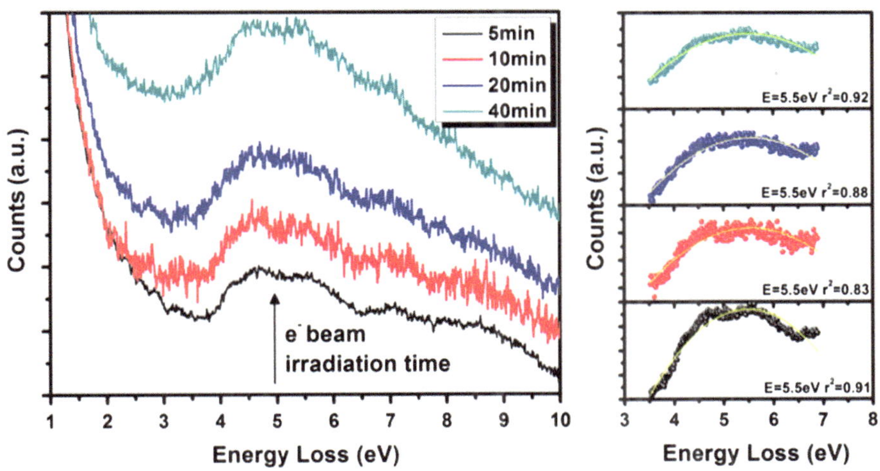

Fig. 2.20 Temporal evolution of the normalized surface-loss Mono-STEM-EELS acquired with the e-beam passing through the same vacuum position adjacent to the e-beam produced bismuth nanoparticle. Reprinted with permission from Ref. [19]

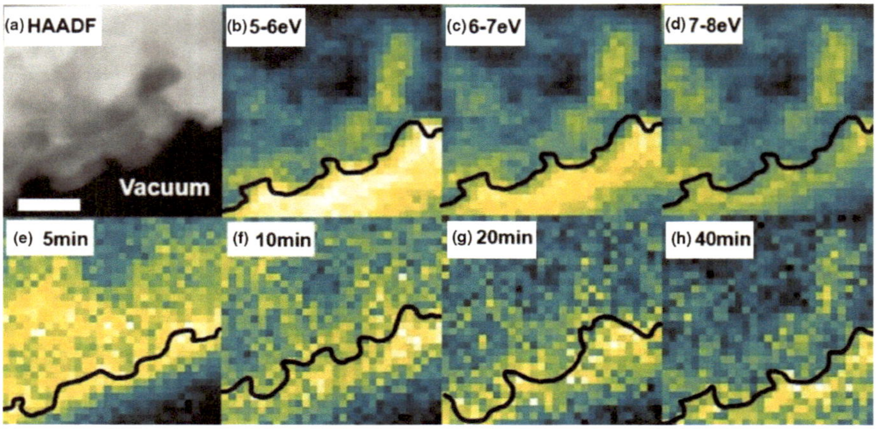

Fig. 2.21 HAADF image of the single-crystalline BiOCl exposed by e-beam for 40 min (**a**, scale bar: 10 nm), the corresponding Mono-STEM-EELS maps extracted at different energy regions (**b–d**), as well as the EELS maps (extracted at 5.5 eV) of the identical specimen region irradiated for different periods (**e–h**). The *solid black lines* show contour of the irradiated specimen. The pixel contrasts in both (**b–d**) and (**e–h**) have been normalized in order to quantitatively show the optical response. Average e-beam dose rate: ca. 0.5 e/A2 s. Reprinted with permission from Ref. [19]

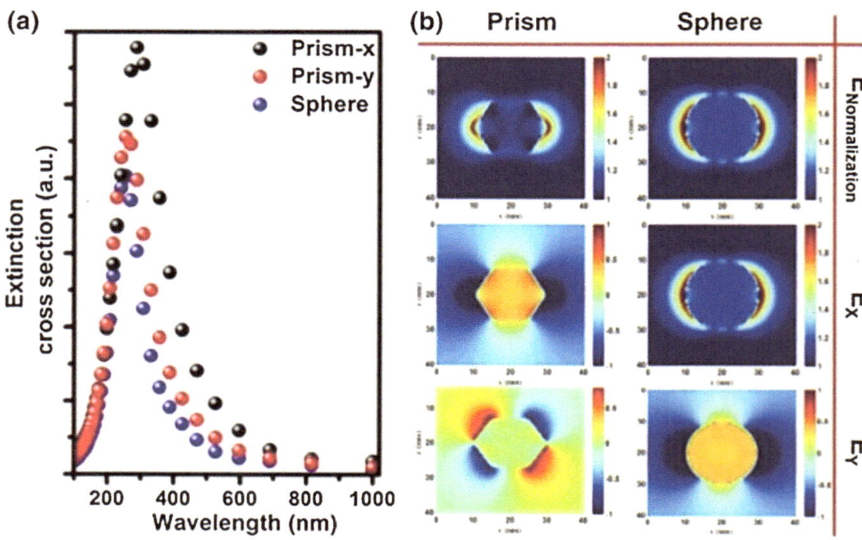

Fig. 2.22 Simulated extinction cross section (**a**) for an individual bismuth nanoparticle with hexagonal prism (edge length = 10 nm, height = 20 nm) and spherical 3-D morphology (diameter = 20 nm), and their corresponding electric filed distributions at 5.5 eV (**b**). Reprinted with permission from Ref. [19]

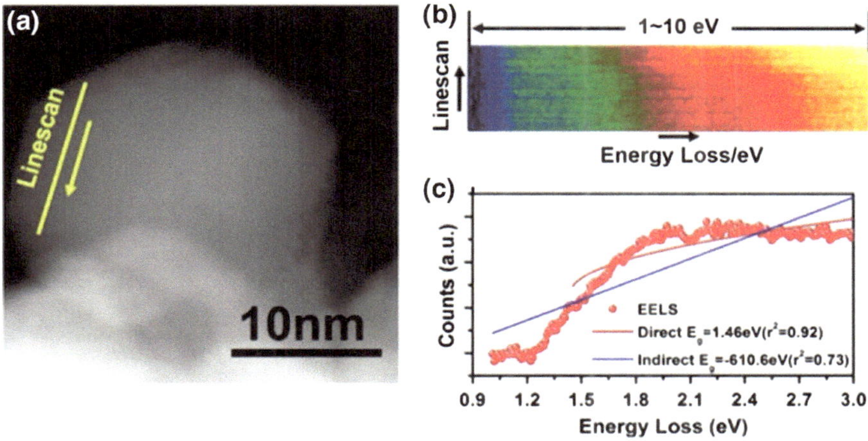

Fig. 2.23 HAADF image of an individual bismuth nanoparticle (**a**, *solid line* and *arrow* indicate the line-scanning region and direction, respectively), and a representative low-loss EELS with Eg determination (**b**, **c**). Reprinted with permission from Ref. [19]

band gap (eV), proportionality constant, and exponent which indicates the characteristic of transition type in a semiconductor (n = 0.5, direct transition; n = 1.5, indirect transition), respectively [26]. As depicted in Fig. 2.23c, when indirect band transition model was adopted, absurd E_g value (indirect $E_g = -610.6$ eV) with quite low fitting coefficient ($r_2 = 0.73$) was obtained. However, reasonable E_g value (direct $E_g = 1.46$ eV) with relatively high fitting coefficient ($r_2 = 0.92$) would have been found if we had used the parabolic equation for direct E_g determination. This indicates the nature of direct band transition in bismuth, which is also in consistent with the DFT calculation (Fig. 2.21, supplementary data, Ref. 20) and previous reports [25]. Since the fitted E_g is strongly replied on the energy window selected [27], the E_g value of e-beam produced single bismuth nanoparticle has been determined statistically at 1.42 ± 0.08 eV by selecting various power law-based EELS background subtractions which take account of the zero-loss contribution (Figs. 2.22 and 2.23). Moreover, supplementary NIR absorption spectrum of commercial bismuth nanoparticles with average size of 40 nm shows the E_g value of ca. 1.58 eV, suggesting a nice agreement with the results obtained via EELS analysis as well. Additionally, based on the line-scanning low-loss EELS results, the computed value for thickness of an individual bismuth nanoparticle was only ca. 6 nm, but the direct transition in Bi NPs with much smaller thickness could facilitate the high efficiency by harvesting entire visible light irradiations.

References

1. Cheng HF, Huang BB, Dai Y (2014) Engineering BiOX (X = Cl, Br, I) nanostructures for highly efficient photocatalytic applications. Nanoscale 6:2009–2026
2. Zhang KL, Liu CM, Huang FQ et al (2006) Study of the electronic structure and photocatalytic activity of the BiOCl photocatalyst. Appl Catal B: Environ 68:125–129
3. Wang DH, Gao GQ, Zhang YW et al (2012) Nanosheet-constructed porous BiOCl with dominant 001 facets for superior photosensitized degradation. Nanoscale 4:7780–7785
4. Jiang J, Zhao K, Xiao XY et al (2012) Synthesis and facet-dependent photoreactivity of BiOCl single-crystalline nanosheets. J Am Chem Soc 134:4473–4476
5. Tian Y, Guo CF, Guo YJ et al (2012) BiOCl nanowire with hierarchical structure and its Raman features. Appl Surf Sci 258:1949–1954
6. Wang CH, Zhang XT, Yuan B et al (2012) Simple route to self-assembled BiOCl networks photocatalyst from nanosheet with exposed (001) facet. Micro & Nano Lett 2:152–154
7. Xiong JY, Cheng G, Li GF et al (2011) Well-crystallized square-like 2D BiOCl nanoplates: mannitol-assisted hydrothermal synthesis and improved visible-light-driven photocatalytic performance. RSC Adv 1:1542–1553
8. Wang CH, Shao CL, Liu YC et al (2008) Photocatalytic properties BiOCl and Bi_2O_3 nanofibers prepared by electrospinning. Scripta Mater 59:332–335
9. Wang YN, Deng KJ, Zhang LZ (2011) Visible light photocatalysis of BiOI and its photocatalytic activity enhancement by in situ ionic liquid modification. J Phys Chem C 115:14300–14308
10. Deng ZT, Tang FQ, Anthony JM (2008) Strong blue photoluminescence from single-crystalline bismuth oxychloride nanoplates. Nanotechnology 19:2123–2131
11. Song JM, Mao CJ, Niu HL et al (2010) Hierarchical structured bismuth oxychlorides: self-assembly from nanoplates to nanoflowersviaa solvothermal route and their photocatalytic properties. CrystEngComm 12:3875–3881
12. Armelao L, Bottaro G, Maccato C et al (2012) Bismuth oxychloride nanoflakes: interplay between composition-structure and optical properties. Dalton T 41:5480–5485
13. Ye LQ, Tian LH, Peng TY et al (2011) Synthesis of highly symmetrical BiOI single-crystal nanosheets and their 001 facet-dependent photoactivity. J Mater Chem 21:12479–12484
14. Xia JX, Yin S, Li HM et al (2011) Improved visible light photocatalytic activity of sphere-like BiOBr hollow and porous structures synthesized via a reactable ionic liquid. Dalton T 40:5249–5258
15. Su DS, Zhang B, Schloegl R (2015) Electron microscopy of solid catalysts—transforming from a challenge to a toolbox. Chem Rev 115:2818–2882
16. Liao HG, Niu KY, Zheng HM (2013) Observation of growth of metal nanoparticles. Chem Commun 49:11720–11727
17. Li LZ, Jokisaari JR, Pan XQ (2015) In situ electron microscopy of ferroelectric domains. MRS Bull 40:53–61
18. Specht P, Ho JC, Xu X, Armitage R, Weber ER, Erni R, Kisielowski C (2005) Band transitions in wurtzite GaN and InN determined by valence electron energy loss spectroscopy. Solid State Commun 135:340–344
19. Chang X, Xie L, Sha W, Lu K, Qi Q, Dong C, Yan X, Gondal MA, Rashid SG, Dai QI, Zhang W, Yang L, Qiao X, Mao L, Wang P (2017) Probing the light harvesting and charge rectification of bismuth nanoparticles behind the promoted photoreactivity onto Bi/BiOCl catalyst by (in-situ) electron microscopy, Appl Catal B 201:495–502
20. Matsukawa M, Ishikawa R, Hisatomi T, Moriya Y, Shibata N, Kubota J, Ikuhara Y, Domen K (2014) Enhancing photocatalytic activity of $LaTiO_2N$ by removal of surface reconstruction layer. Nano Lett 14:1038–1041
21. Xia JY, Tang MT, Chen C, Jin SM, Chen YM (2012) Preparation of α-Bi_2O_3 from bismuth powders through low-temperature oxidation. Trans Nonferrous Met Soc China 22:2289–2294

22. Hu M, Chen JY, Li ZY, Au L, Hartland GV, Li XD, Marquez M, Xia YN (2006) Gold nanostructures: engineering their plasmonic properties for biomedical applications. Chem Soc Rev 35:1084–1094
23. McMahon JM, Schatz GC, Gray SK (2003) Plasmonics in the ultraviolet with the poor metals Al, Ga, In, Sn, Tl, Pb, and Bi. Phys Chem Chem Phys 15:5415–5423
24. Novotny L, Hecht B (2006) Principles of nano-optics. Cambridge University Press
25. Black MR, Lin YM, Cronin SB, Rabin O, Dresselhaus MS (2002) Infrared absorption in bismuth nanowires resulting from quantum confinement. Phy Rev B 65:195417
26. Rafferty B, Brown LM (1998) Direct and indirect transitions in the region of the band gap using electron-energy-loss spectroscopy. Phy Rev B 58:10326–10337
27. Erni R, Browning ND (2005) Valence electron energy-loss spectroscopy in monochromated scanning transmission electron microscopy. Ultramicroscopy 104:176–192

Chapter 3
Band Structure, Band Energy, and Density of States

Abstract This is a short chapter describing crystallographic and structural parameters such as lattice constants, structural characteristics, band structures, densities of states, and atomic charge populations of bismuth oxyhalide crystals. This chapter embodies theoretical calculation and experimental results of the above parameters.

Keywords Lattice constant · Band structure · Density of states · Atomic charge population

Crystallographic, optical, and other structural parameters of BiOX crystals such as lattice constants, structural characteristics, band structures, densities of states, atomic charge populations, and optical parameters can be determined using the first-principles based on the density functional theory. In classic computation methods, the calculations are performed using the commercial software package known as Cambridge Serial Total Energy Package (CASTEP) code of Materials Studio which uses the density functional theory (DFT).

Using the above, Zhao et al. [1] calculated the indirect band gaps of BiOCl, BiOBr, and BiOI crystals and their respective values are 2.50, 2.10, and 1.59 eV. They also reported that the valence band maximum is mainly contributed by O 2p and X np states and the conduction band minimum is decided by Bi 6p states. This is based on the analysis of densities of states and atomic charge populations of BiOX crystals, and the sequence of covalent bond strength between the atoms (Bi–O > Bi–I > Bi–Br > Bi–Cl). Finally, the calculated absorption edges of the absorption coefficients for BiOCl, BiOBr, and BiOI crystals are 355, 448, and 645 nm, respectively, which is in good agreement with their experimental data.

Wang et al. [2] investigated the band offsets at the interface of two photocatalytic materials BiOCl:Bi_2WO_6. They found that the W–O bonded interface was the most stable and the intrinsic interface fails to enhance the charge-carrier separation due to the improper band alignment.

M.A. Gondal et al., *Novel Bismuth-Oxyhalide-Based Materials and Their Applications*, Advanced Structured Materials 76,
DOI 10.1007/978-81-322-3739-6_3

Zhang et al. [3] studied the lattice constants, band structures, total density of states, partial densities of states, and optical absorption spectra of the perfect BiOCl and $BiO_{15/16}Cl$. The formation energies calculated in their work indicates the possible existence of oxygen vacancy in the BiOCl sample and this oxygen vacancy could affect the Bi 6p state and as a result introduce a new electronic state within the forbidden band. Also in the case of $BiO_{15/16}Cl$, they found an additional absorption peak in the visible region at 456 nm compared to BiOCl.

Yang et al. [4] found that compared to pure BiOCl, the band-gap energy of WO_6–BiOCl is reduced by 0.25 eV as a result of the upward shifting of valence band maximum (VBM) and this led to the redshift towards visible light compared to pure BiOCl. All of these results reveal that the tungsten-doped BiOCl is quite beneficial for photocatalytic application.

Huang [5] reported that the valence band width sequentially increases from relaxed BiOCl to BiOI, whereas the conduction band width sequentially decreases from relaxed BiOF to BiOI. They also reported that the estimated dipole moments of BiO_4X_4 polyhedra within the relaxed species are in the order of BiOF > BiOI > BiOBr > BiOCl when Bi 5d states are considered.

Huang and Zhu [6] in another work confirmed that Bi 5d states are instrumental in narrowing the valence bands and widening the conduction bands. They also reported that the increasing tendency of valence band width and the decreasing tendency of conduction band width are sequentially from BiOCl to BiOI.

Zhang et al. [7] investigated the bonding behavior, electronic structure, and visible light absorption of $Mn_xBi_{1-x}OCl$ ($x = 0$, 0.0625, 0.09375, and 0.125) using density functional theory. Their calculated results show that Mn–O covalent bond is stronger than Bi–O covalent bond and also the Fermi level shifts to the middle of forbidden band after Mn doping. Furthermore, Mn 3d states can not only trap the excited electrons, but also possibly trap the photoexcited holes when BiOCl is moderately doped with Mn. Moreover, their calculated optical absorption spectra exhibit an obvious redshift in the spectral absorption edge which is in agreement with the experimental results.

References

1. Zhao LJ, Zhang XC, Fan CM et al (2012) First-principles study on the structural, electronic and optical properties of BiOX(X = Cl, Br, I) crystals. Phys B 407:3364–3370
2. Wang WC, Yang WJ, Chen R et al (2012) Investigation of band offsets of interface BiOCl: Bi_2WO_6 a first-principles study. Phys Chem Chem Phys 14:2450–2454
3. Zhang XC, Zhao LJ, Fan CM et al (2012) Effects of oxygen vacancy on the electronic structure and absorption spectra of bismuth oxychloride. Comput Mater Sci 61:180–184
4. Yang WJ, Wen YW, Chen R et al (2014) Study of structural, electronic and optical properties of tungsten doped bismuth oxychloride by DFT calculations. Phys Chem Chem Phys 16:21349–21355
5. Huang WL (2008) Electronic structures and optical properties of BiOX (X = F, Cl, Br, I) via DFT calculations. J Comput Chem 12:1882–1891

6. Huang WL, Zhu QS (2008) DFT calculations on the electronic structures of BiOX (X = F, Cl, Br, I) photocatalysts with and without semicore Bi 5d states. J Comput Chem 2:183–190
7. Zhang XC, Zhao LJ, Fan CM et al (2012) First-principles investigation of impurity concentration influence on bonding behavior, electronic structure and visible light absorption for Mn-doped BiOCl photocatalyst. Phys B 407:4416–4424

Chapter 4
Optical Properties of Bismuth Oxyhalides

Abstract This chapter is especially dedicated to the optical properties of bismuth oxyhalides. The absorption and reflectance spectra to study the spectral absorption and the band-gap energies of the material, photoluminescence spectra to study the recombination of photoinduced charge-carriers are presented.

Keywords Reflectance spectra · Absorption spectra · Kubelka-Munk function · Photoluminescence spectra

4.1 Optical Properties of Semiconductors

Semiconductors are characterized by fully occupied valence band and unoccupied conduction band. Energy band in general is the fusion of completely degenerated molecular energy levels arising from the mixing of multitudes of molecular orbital. In a solid there are several bands of energies, but for any practical point of view we consider only the uppermost bands known as valence band and conduction band. The electrons in the inner orbital of an atom are less influenced by the surrounding atoms; and hence they form discrete energy levels, on the other hand the outer electrons or the valence electrons can move from atom to atom, but only up to the top of the valence band. In order for the electrons to be available for conduction (or for any chemical reaction) in semiconductor, they have to be promoted to the conduction band electrical, thermal, or optical means. So for any application of semiconductor material, band-gap energy, which is the energy difference between the top of the valence band and the conduction band, is a crucial factor.

When we consider the optical characteristics of the semiconducting material, the classification of this material as direct and indirect band-gap material is very important. In the case of direct band-gap material, the momentum of electrons in the highest level of the valence band and the lowest level of conduction band is the same, whereas in the indirect band-gap material the electrons in the lowest conduction band have higher or lower momentum than the electrons in the highest level of the valence band. When the photon of energy is equal to or more than the band

M.A. Gondal et al., *Novel Bismuth-Oxyhalide-Based Materials and Their Applications*, Advanced Structured Materials 76,
DOI 10.1007/978-81-322-3739-6_4

gap, energy of the material strikes and electron hole pair is produced; in the case of direct band-gap semiconductor, this process is quite easier as the electrons do not need to undergo any momentum change, whereas in the case of indirect band-gap semiconductors, the electron has to undergo a significant change in the momentum for a given photon energy in order to produce electron hole pair. This conservation of momentum is possible for such electrons only through the intervention of phonon from the lattice vibration in addition to the exciting photons. Due to this three particle interaction (electron, photon, and phonon), the rate of generation of electron hole pair in indirect semiconductors is much slower than the direct semiconductors. This difference between the direct and indirect semiconductor material is true for the reverse process, where the electrons and the holes are recombined to form photons, where the recombination of electron hole pair is easier in the case of direct band-gap material than the indirect semiconductors.

4.2 Absorption/Reflectance Spectra of Bismuth Oxyhalides

Two optical characterizations, absorption spectroscopy, and the photoluminescence spectroscopy are usually employed to study the nature of the semiconducting material, where the former is used to find the band-gap energy of the material and the latter is used to study the electron hole recombination process. As long as the photon energy of the exciting photon is lower than the band-gap energy of the semiconducting material, the electrons remain in the valence band and the moment the photon energy is equal to the band-gap energy; the electrons absorb the photon energy and goes to the conduction band by leaving a positively charged hole in the valence band. Thus, electron hole pair is generated in the semiconductor. The wavelength (or photon energy) where the absorption just begins is equal to the band-gap energy of the material.

Diffuse reflectance spectrum is used to find the band-gap energy of the material. When the light falls on the semiconductor (powder sample), the light is reflected in all directions due to the varied shapes of the particles and this reflected light is collected by an integrating sphere and directed to the detector. The specular reflection takes place in the case of highly smooth reflecting surface like mirror, where the law of reflection is clearly evident. In the case of diffuse reflection, the particles are differently shaped and oriented and hence the reflected beam goes into different directions by obeying law of reflection for individual particles. The wavelength at which the absorption takes place the diffuse reflectance signal is less, like in the case of transmission spectrum.

Kubelka-Munk Function $F(R)$ is the representation of the absorption spectrum of the sample from the diffuse reflectance spectrum through the following transformation:

$$F(R) = \frac{(1-R)^2}{2R} = K/S,$$

where R is the absolute reflectance taken from the diffuse reflectance spectrum and K and S are the absorption and scattering coefficients, respectively. From the above equation, it is clear that when the Kubelka-Munk function is plotted against the wavelength of the incident light, $F(R)$ is a good representation of the absorption data because the scattering coefficient is less dependent on the wavelength. For determining the band gap of the semiconductor for the absorption data we use Tauc Plot, where we plot energy in the x-axis and the (energy * absorption coefficient)^n in the y-axis and the extrapolation of the linear portion of the curve yield the band-gap energy of the material.

Zhang et al. [1] synthesized bismuth oxyhalides by adding $Bi(NO_3)_3$ into an EG solution containing stoichiometric amounts of KCl, NaBr, and KI and from these mixtures BiOX (X = Cl, Br, and I) were prepared by a solvothermal process. Reflectance spectra of BiOX (X = Cl, Br and I) indicates that the absorption edge of BiOX (X = Cl, Br and I) occurs at 370, 440, and 670 nm; these results are consistent with the white, light yellow, and red colors of BiOCl, BiOBr, and BiOI, respectively. In their study, BiOX as a crystalline semiconductor, the optical absorption near the band edge could be determined by following this theoretical equation:

$$\alpha h\nu = A(h\nu - E_g)^{(n/2)},$$

where α, ν, A, and E_g are the absorption coefficient, light frequency, a constant, and band-gap energy, respectively. Among them, n depends on the characteristic transitions in a semiconductor. (i.e., n = 1 for direct transition or n = 4 for indirect transition). For BiOX, as an indirect band-gap semiconductor, the value of n is 4 for the indirect transition [2]. Thus, the band-gap energies (E_g) of the BiOX samples could be estimated by extrapolating the straight portion of $(\alpha h\nu)^{1/2}$ versus the photon energy ($h\nu$). A good approximation of the band-gap energies for the BiOX is obtained by extrapolating the tangent to the X-axis. Thus, the band-gap energies are estimated to be 3.22, 2.64, and 1.77 eV, respectively, for BiOCl, BiOBr, and BiOI as shown in Figs. 4.1 and 4.2.

4.3 Photoluminescence Spectra of Bismuth Oxyhalides

Photoluminescence spectroscopy is an important characterization tool for the semiconductors. As mentioned earlier, the energy bands in a semiconductor is like the atomic and molecular energy levels capable of absorbing and emitting UV–VIS radiation under suitable conditions. In a semiconductor when the photon energy of the excitation source is more than the band-gap energy, the energy is absorbed by

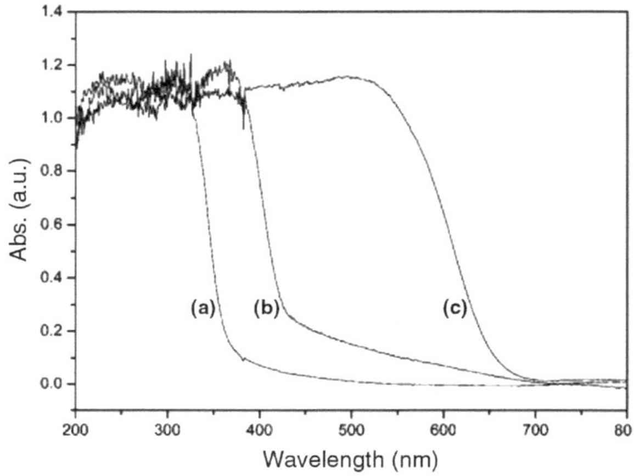

Fig. 4.1 UV–VIS diffuse reflectance spectra (DRS) of BiOX (X = Cl, Br, I) powders. **a** BiOCl, **b** BiOBr, and **c** BiOI. Reprinted with permission from Ref. [1]

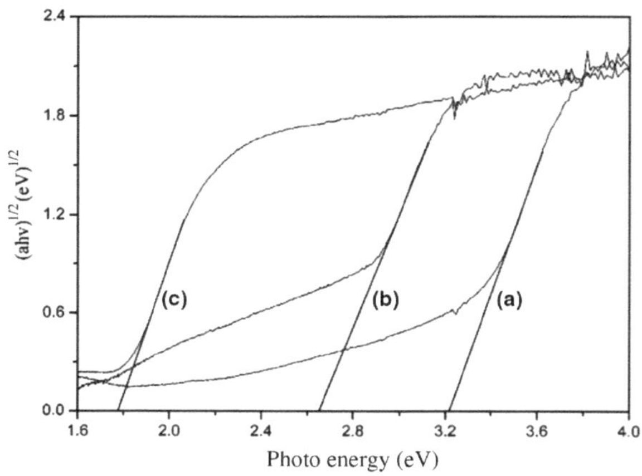

Fig. 4.2 Plots of $(\alpha h\nu)^{1/2}$ versus photon energy ($h\nu$) for BiOX (X = Cl, Br, I) powders. **a** BiOCl, **b** BiOBr, and **c** BiOI. Reprinted with permission from Ref. [1]

the electrons in the valence band and with this excess energy electrons move to conduction band leaving a positively charged hole in the valence band. These generated electron hole pairs (excitons) undergo various recombination processes and in some processes the energy is released in the form of light radiation, which is a key for the photoluminescence spectroscopy. The life time of the photogenerated excitons depends on many characteristics of the semiconductor like the nature of crystal lattice, type and concentration of the impurities present, and crystallographic defects and dislocations. In addition to the excitonic recombination, other radiative processes that lead up to the photoluminescence signal are band -to-band recombination, due to excitons binding to the donors, free electron binding to the acceptors, and also donor acceptor recombination.

The solution of the Schrodinger equation of the electron hole pair is similar to that of hydrogen atom and is given by $E_n = -\frac{\mu e^4}{2h^2\varepsilon^2 n^2}$, where μ is the reduced mass of the electron hole pair, ε is the dielectric constant, and n is the quantum number of the exciton state. Due to the coulombic attraction between electron and hole, the radiative energy due to this process is less than the band-gap energy and also the transitions are possible within the electron hole pair energy levels described above. Like in the case of electron hole pair recombination, the similar treatment can be applied to other radiative processes such as an exciton bound to donor, acceptor, or donor–acceptor pair.

The concentrations and lifetimes of non-equilibrium electrons and holes are equal when no obvious defects are in the samples. The average lifetime (τ_{av}) of the non-equilibrium carriers [3] is given as

$$\tau_{av} = a_1\tau_1 + a_2\tau_2/a_1 + a_2,$$

where τ_1 is average lifetime for excitonic state (s), a_1 is amplitude of bandedge-excitonic state (a.u.), τ_2 is average lifetime for trapping state emission (s), and a_2 is amplitude of trapping state emission (a.u.).

Deng et al. [4] adopted a solution route of hydrogen peroxide oxidation of bulk metal bismuth particles in the presence of sodium chloride, and BiOCl nanoplates were synthesized by hydrothermal treatment. In their study, the emission spectrum shows a peak position at 455 nm with an excitation wavelength of 364 nm as shown in Fig. 4.3, and also the photographs of the BiOCl nanoplate colloidal solution under a 365 nm UV-lamp that obviously show the strong blue photoluminescence of the nanoplate solution. It is found that by comparison with a standard sample (2-aminopyridine ($\geq$99%) in 0.1 M H_2SO_4, Φ_{PL} = 60%), the photoluminescence quantum yield (Φ_{PL}) was estimated to be ~0.4. In this research, the room temperature time–decay curve of the QDs with an emission at 455 nm fits commendable to a bi-exponential function as in Fig. 4.4. And then the average decay time (τ) was calculated by a fast part of 1.949 ns (61.42%) and a slow part of 8.448 ns (38.58%) is 6.7 ns, where the fast process equivalents to the

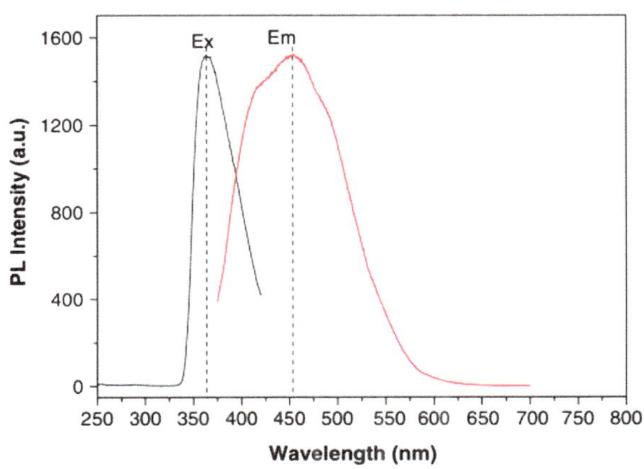

Fig. 4.3 Photoluminescence emission spectra of BiOCl. Reprinted with permission from Ref. [4]

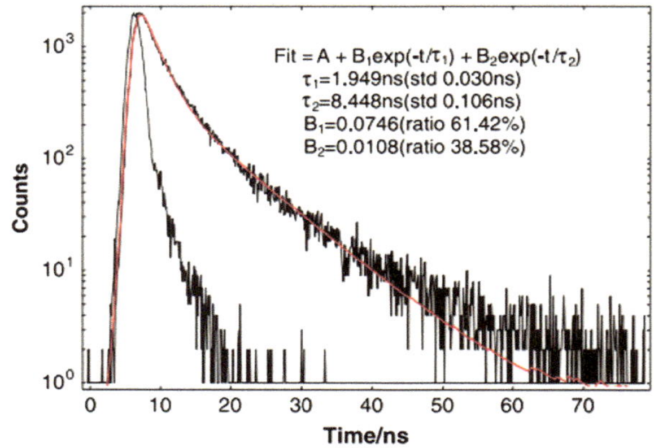

Fig. 4.4 Luminescence decay curves of BiOCl. Reprinted with permission from Ref. [4]

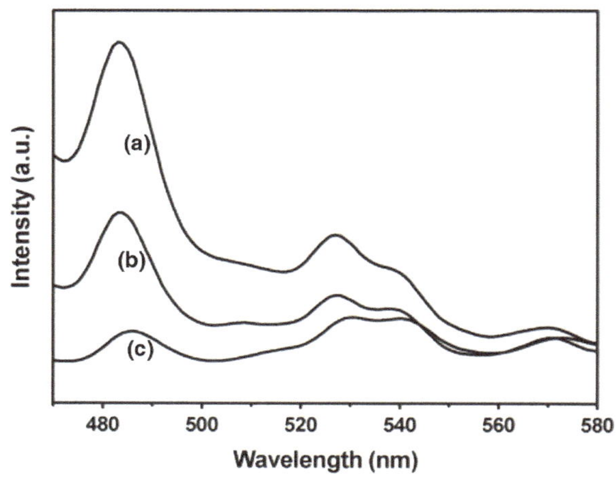

Fig. 4.5 The PL spectra of the BiOI with different morphological structure. **a** nanoplates; **b** hollow microspheres; and **c** flower-like microspheres. Reprinted with permission from Ref. [5]

electron–phonon interactions, and the slow process amounts to the energy relaxation of the hot lattice by phonon–phonon interactions to the ambient medium.

Ren et al. [5] added $Bi(NO_3)_3$ into a water–ethanol mixed solution with the assistance of PVP and citric acid at 70 °C, and then BiOI was prepared through precipitation route. The emission spectrum shows two main emission peaks at 485 nm and 526 nm with excitation at 255 nm as shown in the Fig. 4.5. In this study, it can be obtained clearly that different structures corresponds to different PL intensities: the nanoplates were highest, the hollow microspheres were second and better than flower-like microspheres.

References

1. Zhang X, Ai ZH, Jia FL (2008) Generalized one-pot synthesis, characterization, and photocatalytic activity of hierarchical BiOX (X = Cl, Br, I) nanoplate microspheres. J Phys Chem C 112:747–753
2. Zhang KL, Liu CM, Huang FQ (2006) Study of the electronic structure and photocatalytic activity of the BiOCl photocatalyst. Appl Catal B Environ 68:125–129
3. Singh R, Pal B (2013) Study of excited charge carrier's lifetime for the observed photoluminescence and photocatalytic activity of CdS nanostructures of different shapes. J Mol Catal A Chem 371:77–85
4. Deng ZT, Tang FQ, Anthony JM (2008) Strong blue photoluminescence from single-crystalline bismuth oxychloride nanoplates. Nanotechnology 19:2123–2131
5. Ren KX, Zhang K, Liu J (2012) Controllable synthesis of hollow/flower-like BiOI microspheres and highly efficient adsorption and photocatalytic activity. CrystEngComm 14:4384–4390

Chapter 5
Preparation Techniques and Crystal Growth Processes

Abstract This chapter presents different preparation methods for bismuth oxyhalides and their relative merits, characteristics, and applicability. The method of preparation of different variants of bismuth oxyhalides includes various hydrolysis methods, preparation by heat treatment, vapor-phase synthesis, hydrothermal preparation, solvothermal preparation, and special halide source-induced preparation.

Keywords Hydrolysis method · Preparation by heat treatment · Vapor-phase synthesis · Hydrothermal preparation · Solvothermal preparation

Synthesis of bismuth oxyhalides BiOX (X = Cl, Br, I) by novel techniques is still a topic of great research interest, although the discovery of these materials is dated back to nineteenth century [1]. With the help of the modern analytical instrumentation, the BiOX so synthesized by various processes were characterized for their relative microstructures, self-assembly processes, and other merits. In the last 10 years, we have witnessed a tremendous progress in the controlled synthesis processes of BiOX in terms of their morphology, crystallography, and chemical interfacial properties. In this chapter, we focus on the most successful and proven preparation techniques of bismuth oxyhalides.

5.1 Preparation by Hydrolysis (Precipitation) Method

5.1.1 Bi^{3+}–X^{-} (X = Cl, Br, I) Aqueous System

The Bi^{3+}–X^{-} (X = Cl, Br, I) aqueous system has been widely adopted to synthesize BiOX (X = Cl, Br, I) compounds of high purity. The typical preparation process involves the following steps: (i) dissolving the bismuth salts like $BiCl_3$, $Bi(NO_3)_3$, or bismuth oxides in aqueous solutions of hydrogen halides (or metal halides) to form the Bi^{3+}–X^{-} system of very low pH values (usually less than 1); (ii) adding

M.A. Gondal et al., *Novel Bismuth-Oxyhalide-Based Materials and Their Applications*, Advanced Structured Materials 76,
DOI 10.1007/978-81-322-3739-6_5

alkali solutions like H_2O, NH_4OH, Na_2CO_3 into the Bi^{3+}–X^- system in order to maintain the pH of the solution around 2–3; and (iii) collect and wash the precipitation until the solution is completely free from X^- ions and further the precipitate is dried to get the final BiOX product.

Zheng et al. [2] investigated the residue phase in the process of distillation of $BiCl_3$–HCl–H_2O system. It was found that the required concentration ratio ([Cl]/[Bi]) decreased with the increase of temperature, and the phases of $BiCl_3{\cdot}0.12H_2O$ and $BiCl_3$ were found at 140 °C. When the temperature was over 160 °C, concentration ratio ([Cl]/[Bi]) was less than 3:1, indicating that the coexistence of $BiCl_3$ and BiOCl. Further increase in temperature results in the increase of mass ratio of BiOCl phase. Their theoretical calculation shows that the hydrolysis could take place over 128 °C.

Wang et al. [3] studied the thermodynamic equilibrium of bismuth hydrometallurgy in chloride solution. The $lgc(Bi)_T$-pH diagrams of BiOCl, Bi_2O_3, $Bi(OH)_3$, and BiOOH in the aqueous solution containing chloride ions is shown in Fig. 5.1. It can be found that Bi_2O_3 and BiOOH occupy the widest and narrowest stable region, respectively. Also the obvious effect of $c_0(Bi^{3+})$ and $c_0(Cl^-)$ on the equilibrium of BiOCl-solution is found in the diagram. When reducing $c_0(Cl^-)$ from 3.0 to 0.3 mol L^{-1}, the hydrolysis pH for BiOCl formation moves from 3 to 2. This makes it quite clear that the pH value of the solution has a major role in obtaining high purity BiOCl and should be very carefully controlled. Moreover, low $c_0(Bi^{3+})$ and $c_0(Cl^-)$ in aqueous solution should also be maintained as the pH value for the formation of BiOCl is lower than that of Bi_2O_3, and it is independent on the pH value of the solution.

Wosylus et al. [4] carried out time-dependent in situ X-ray powder diffraction and time and temperature-controlled TG-MS (should be expanded) and their experiments revealed that the intermediate hydrate $BiCl_3{\cdot}H_2O$ is formed in the hydrolysis process of transforming $BiCl_3$ (in water vapor) to BiOCl (as illustrated in Scheme 5.1) [5]. The time-dependent X-ray powder diffraction results indicate that

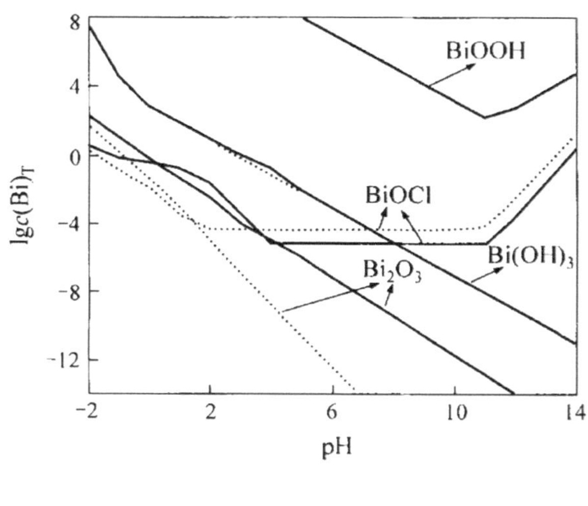

Fig. 5.1 $lgc(Bi)_T$-pH diagram of Bi(III)–Cl–H_2O system (*solid line* and *dotted line* represents $c_0(Cl^-)$ of 3.0 and 0.3 mol L^{-1}, respectively). Should we put lgc as $\log[c(Bi)_T]$. Reprinted with permission from Ref. [3]

Scheme 5.1 Reaction of $BiCl_3$ with H_2O (g)

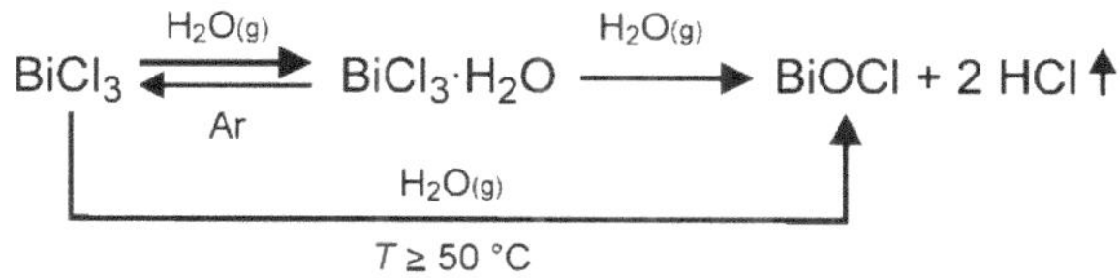

$BiCl_3$ reacts with the moisture in the ambient air to give intermediate $BiCl_3{\cdot}H_2O$ followed by the formation of BiOCl. By tracing the mass changes of the $BiCl_3$ (purified by sublimation) as a function of reaction time, it was found that 4.6 and 3.8wt% mass was gained at 27 and 40 °C, respectively, at the initial stage, demonstrating the formation of the intermediate hydrate. The subsequent linear mass loss indicates the formation of BiOCl. The XRD patterns of the sample at 27 °C shows the maximum weight gain, indicating the existence of $BiCl_3{\cdot}H_2O$ phase as the main component, compared to other two phases of $BiCl_3$ and BiOCl. The TG-MS results clearly shows a simultaneous increasing of ion current at $m/z = 18$ (H_2O^+) with onset at about 50 °C, and this indicates the hydrolysis of hydrate (i.e. $BiCl_3{\cdot}H_2O$). Also several $BiCl_x^+$ ($x = 1, 2, 3$) species in the MS could be observed at about 250 °C, where the sublimation of $BiCl_3$ starts.

5.1.2 *Other Hydrolysis Processes*

Novokreshchenova et al. [6] developed a novel preparation method for BiOCl as a product of the reaction between Bismuth (III) trihydrate oxohydroxonitrate ($[Bi_6O_5(OH)_3](NO_3)_5{\cdot}3H_2O$) and NH_4Cl at ambient temperatures (23 and 60 °C). The precursor of Bismuth trihydrate oxohydroxonitrate was prepared by adding ammonium carbonate into the bismuth-containing nitric acid solution. The Bismuth oxochloride formation can be described by the following chemical equation:

$$[Bi_6O_5(OH)_3](NO_3)_5{\cdot}3H_2O + 6NH_4Cl \rightarrow 6BiOCl + 5NH_4NO_3 + NH_4OH + 4H_2O.$$

The TG-DTA (should be expanded) data of the as-prepared Bismuth trihydrate oxohydroxonitrate provided in their report is shown in Fig. 5.2. It is clear from Fig. 5.2 that the dehydroxilation process starts at the temperature of 270 °C, followed by five endothermic peaks centered at the temperatures 330, 375, 440, 520, and 600 °C, which are attributed to dehydroxilation complement, five nitrate ions destruction and Bi_2O_3 phase formation, respectively, where the endothermic peak at 730 °C is attributed to the conversion of α-Bi_2O_3 into δ-Bi_2O_3. Their results also indicate that the extraction level of Bismuth into BiOCl is 99.99 and 99.94% for the use of NH_4Cl and HCl solutions, respectively. The following reaction mechanism can be proposed based on their findings.

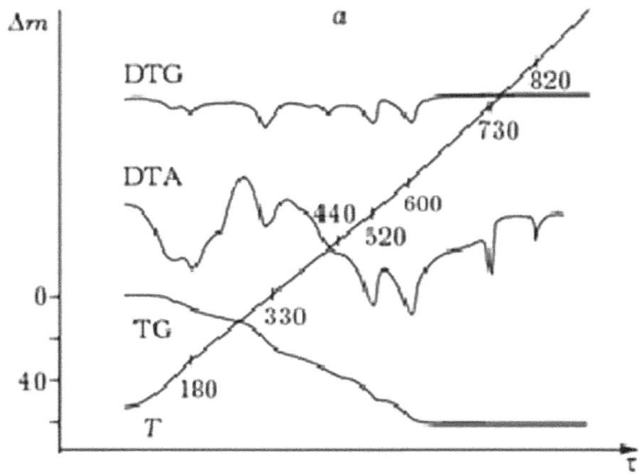

Fig. 5.2 TG-DTA results of bismuth oxohydroxonitrate (heating rate 10 °C/min, specimen mass: 300 mg). Reprinted with permission from Ref. [6]

$$[Bi_6O_5(OH)_3](NO_3)_5{\cdot}3H_2O + 6HCl \rightarrow 6BiOCl + 5HNO_3 + 5H_2O.$$

Ye et al. [7] developed a novel hydrolysis process for the preparation of BiOCl using a special molecular precursor $Bi_n(Tu)_xCl_{3n}$ (Tu = thiourea) as the starting material. This molecular precursor $Bi_n(Tu)_xCl_{3n}$ has a flower-like hierarchical structure as shown in Fig. 5.3. In this process, BiOCl produced through hydrolysis of $Bi_n(Tu)_xCl_{3n}$ takes very short duration, which is as low as 30 s. The molecular precursor is also prepared quite easily from $BiCl_3$ and Thiourea (Tu) according to the following chemical reaction:

$$nBi^{3+} + m\mathrm{Tu} \rightarrow Bi_n(\mathrm{Tu})_xCl_{3n}.$$

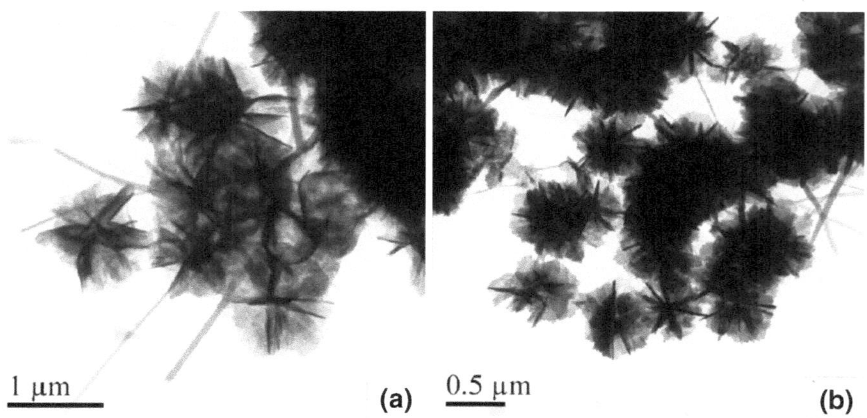

Fig. 5.3 TEM images of $Bi_n(Tu)_xCl_{3n}$ samples prepared at different molar ratios ($BiCl_3$/Tu) at 1:1 (**a**) and 1:3 (**b**). Reprinted with permission from Ref. [7]

5.1.3 Precipitation Preparation of BiOF

To date, only a few studies on the preparation of BiOF compound have been reported. Although the reaction mechanism of BiOF is less understood and need further investigation, to the best of our knowledge, the precipitation method which involves the reaction between Bi^{3+} and F^{-} in aqueous solution in the presence of alkali solution is now considered as the main chemical reaction for the preparation of BiOF. In this preparation method, NH_4F is used as a florid source, because NH_4F is much safer to use than HF and as it is in the solid form, it is easier to handle.

Lee et al. and Kang et al. [8, 9] prepared BiOF coating layer on spinel $Li_{1.1}Al_{0.05}Mn_{1.85}O_4$ using $Bi(NO_3)_3$ and NH_4F as the staring materials. The $BiO_{0.5}F_2$ layer was first obtained by introducing NH_4OH into the mixture of $Bi(NO_3)_3$ salt (dissolved nitric acid) and NH_4F in aqueous solution. The final product, BiOF, was obtained by heating as-precipitated $BiO_{0.5}F_2$ at 400–450 °C temperature for 5 h. The heat treatment in air could lead to the evaporation of fluorine, at the same time the oxygen is gained through the following reaction:

$$4BiO_{0.5}F_2 + O_2 \rightarrow 4BiOF + 2F_2\uparrow.$$

However, further increase of temperature may not favor the formation of BiOF, because at the temperature of 750 °C Bi_2O_3 is produced through the following reaction,

$$4BiOF + O_2 \rightarrow 2Bi_2O_3 + 2F_2\uparrow.$$

Later on, the synthesis of BiOF was successfully achieved at 400 °C by Su et al. [10] using similar preparation scheme. In their work, the XRD patterns of the as-precipitated product, calcined at different temperatures showed the phases of BiF_3 and Bi_2O_3 after annealing with 200 °C, and the phase of $BiO_{1.18}F_{0.64}$ was observed after annealing the precipitate at 500 °C. It is interesting to point out that a few diffraction peaks could not be indexed in the XRD patterns of the as-precipitated product calcined at 300 °C, but some BiOF diffraction peaks, such as (001), (110) and (102) can already be seen.

As a matter of fact, BiOF crystal could also be obtained through the reaction of Bi^{3+} and F^{-} without any further heat treatment. The first example is the report by Wang et al. [11] which shows that the high purity of BiOF can be synthesized using $Bi(NO_3)_3$ and NaF as the raw materials, in the presence of diethanolamine (DEA). Thus, the DEA molecules were considered to play a vital role in the formation of BiOF crystal and the authors deduced that the DEA could facilitate the hydrolysis reaction of Bi^{3+} as it is the characteristic of soft base as shown in Fig. 5.4. Also due to the high stability constants (lgK) of DEA-Bi, it was inferred that DEA leads to the formation of some coordination complex by coordinating with Bi^{3+}, which participates to form BiOF as the end product.

Another example of this method of preparation of BiOF was reported by Bervas et al. [12]. In their method of preparation, Bi_2O_3 was dissolved in NH_4F solution

Fig. 5.4 Chemical structure of diethanolamine (DEA)

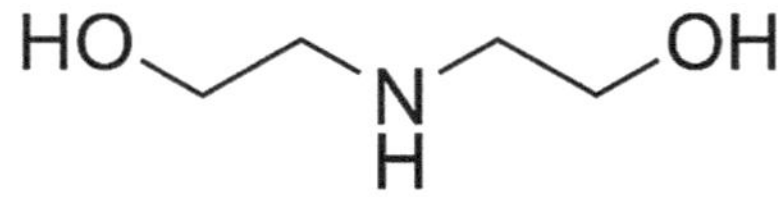

with the molar ratio of $[NH_4F]/[Bi_2O_3]$ between 20 and 100 and stirred for 3 h at room temperature. The power so obtained by this was dried under vacuum at a temperature of 70 °C for a day to obtain BiOF compound. When the molar ratio of $[NH_4F]/[Bi_2O_3]$ is around 10, the BiF_3 and some others unknown phases appeared in the XRD pattern and when the molar ratio is as high as 180, substantial formation of α-NH_4BiF_4 was detected. The whole chemical process can be explained through the following series of chemical reaction:

$$Bi_2O_3 + 2NH_4F \rightarrow 2BiOF + 2NH_3 + H_2O$$

$$Bi_2O_3 + 8NH_4F \rightarrow 2NH_4BiF_4 + 6NH_3 + 3H_2O$$

$$BiOF + 3NH_4F \rightarrow NH_4BiF_4 + 2NH_3 + H_2O.$$

5.2 Preparation by Heat Treatment

There have been many reports about the preparation of bismuth oxyhalides by heat treating the precursor. Ye et al. [13] synthesized highly symmetrical BiOI single-crystal nanosheets by direct annealing BiI_3 in air at a temperature of 260 °C for 16 h and at a temperature of 350 °C for 3 h with heating rate of 5 K/min. The XRD results indicated that the intensity of (001) peak of BiOI, prepared by this method is much higher than that of the BiOI sample prepared by traditional method, implying the presence of higher percentage of {001} facets of BiOI crystals by this method of preparation. Figure 5.5 is the depiction of TG-DSC curves from the work of Ye et al., which clearly shows three apparent weight loss regions corresponding to the following reactions: in 300–420 °C weight loss region, the main reaction, $2BiI_3 + O_2 \rightarrow 2BiOI + 2I_2\uparrow$ (theoretical weight loss of 40.50%) and by-product reaction, $2BiI_3 \leftarrow \rightarrow 2Bi + 3I_2\uparrow$ take place, in the 420–610 °C weight loss region the reaction $5BiOI + O_2 \rightarrow Bi_5O_7I + 2I_2\uparrow$ (theoretical weight loss of 28.89%) takes place and finally in 610–750 °C region, $2\ Bi_5O_7I + 1/2O_2/5Bi_2O_3 + I2\uparrow$ (theoretical weight loss of 6.61%) reaction takes place.

TG-DSC results also indicate that the phase of BiOI is maintained till 350 °C, which is consistent with the work of Yu et al. [14], investigating the effect of heat treatment on BiOI phase. It was also found that the crystals of Bi_5O_7I, (Bi_5O_7I + α-Bi_2O_3), and α-Bi_2O_3 could be obtained at temperature regions of 450–650, 750, and 850 °C, respectively. Also Ueda et al. [15] reported the synthesis of BiOCl by heating $BiCl_3$ in the temperature range of 923–1123 °K for 20 h.

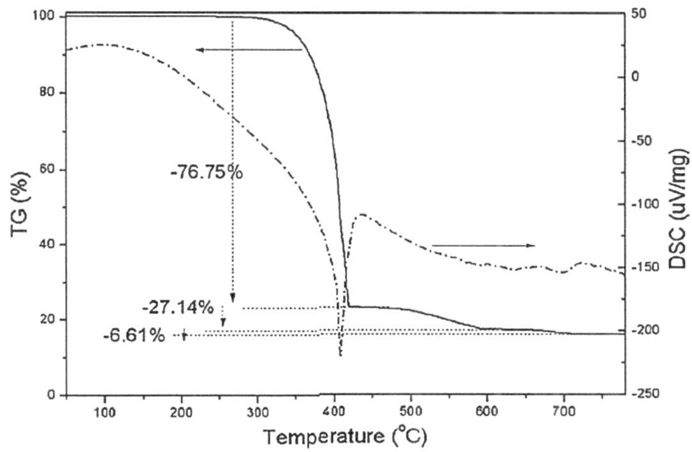

Fig. 5.5 TG-DSC curves of BiI_3 in air (heating rate of 10 °C min^{-1}, air flow of 20 ml L^{-1}) Reprinted with permission from Ref. [13]

5.3 Vapor-Phase Synthesis

Bismuth oxide halide such as BiOCl or BiOI crystals can also be prepared through the vapor-phase synthesis route. In early 1973, Silvestri et al. [16] found that the BiOCl crystals were readily formed when the vapor from $BiCl_3$ kept in the quarts boat inside a reaction chamber is allowed to mix with H_2O vapor by passing N_2O or helium gas to a reaction chamber maintained at a temperature range of 700–800 °C as in Fig. 5.6.

Peng et al. [5] developed a novel synthesis route for BiOCl nanostructure with the controllable morphology by direct vapor-phase through vapor transportation of $AuCl_3$/Bi mixture or $BiCl_3$ at 250 °C temperature as shown in Fig. 5.7. In this process, either $AuCl_3$/Bi or $BiCl_3$ can be used as the starting materials for the

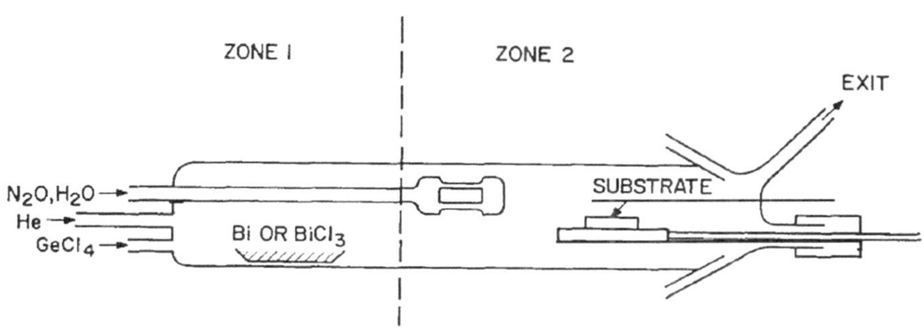

Fig. 5.6 Pertinent features of the experimental system for $Bi_{12}GeO_{20}$, γ-Bi_2O_3 and BiOCl preparation. Reprinted with permission from Ref. [16]

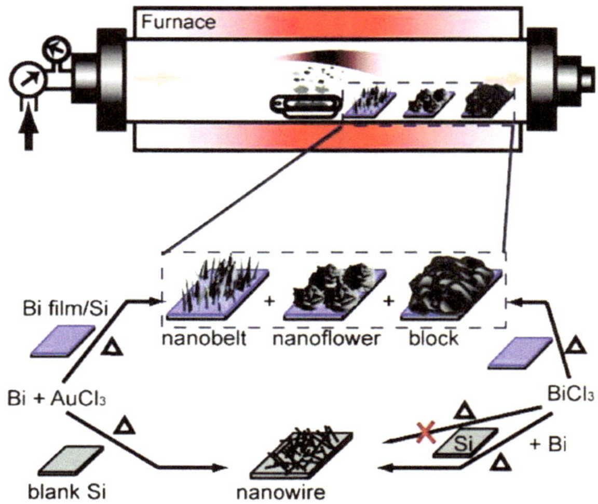

Fig. 5.7 Chemical vapor transport approach as utilized for the controlled growth of BiOCl nanostructures. Reprinted with permission from Ref. [5]

preparation of BiOCl. For the $AuCl_3$/Bi preparation route, the following reaction takes place in the 160–250 °C temperature range:

$$AuCl_3 \rightarrow AuCl + Cl_2,\ Cl_2 + Bi \rightarrow BiCl_3.$$

The vaporized $BiCl_3$, so formed from the above reaction, carried by the stream of inert gas is finally hydrolyzed to BiOCl when the melting point of $BiCl_3$ reached at 234 °C and the chemical reaction is shown below:

$$BiCl_3 + H_2O \rightarrow BiOCl + 2HCl.$$

BiOCl can also be prepared in an evacuated system in the absence of O_2 or water vapor. Keramidas et al. [17] prepared BiOCl thin plates by heating the mixture of high purity Bi_2O_3 and $BiCl_3$ in stoichiometric proportion in a vacuum system (10^{-6} Torr) at the temperature of 820 °C, followed by gradually cooling it down to room temperature.

Similar method was adopted by Ganesha et al. [18] to prepare BiOI crystal using BiI_3 and Bi_2O_3 of very high purity (99.999%) as the starting materials and annealing it at 450 °C for 4 days.

Schuisky and Hårsta [19] theoretically plotted the stability of chemical vapor deposition diagram for the BiI_3–O_2 and BiI_3–H_2O system by employing free-energy minimization technique (computer program of EKVICALC) and the plot is depicted in Fig. 5.8, and the growth of BiOI crystal by CVD under different oxygen sources (i.e., O_2 and H_2O) was studied.

In the diagram for BiI_3–O_2 system, it can be noticed that at a temperature below 200 °C and the $\log\{[O_2]/[BiI_3]\}$ value up to 2, there is a small area which corresponds to the stable region for the solid BiOI crystal. Although a large stability area of BiOI phase is noticed in the diagram for the BiI_3–H_2O system, the BiI_3 phase is

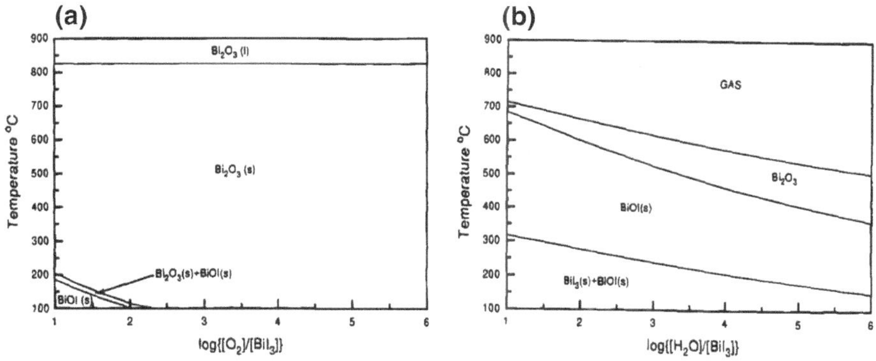

Fig. 5.8 Calculated CVD stability diagram for the BiI_3–O_2 (**a**) and BiI_3–H_2O system (**b**), a total pressure of 10 Torr was chosen for both systems for the calculation. Reprinted with permission from Ref. [19]

Fig. 5.9 Experimental CVD stability diagram for the BiI_3–O_2 system. Reprinted with permission from Ref. [19]

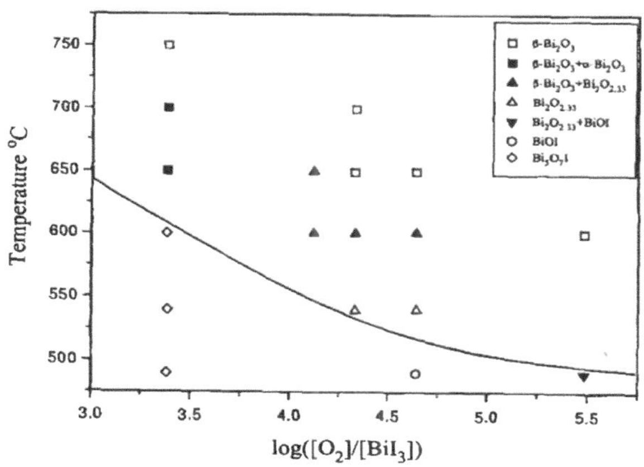

still present at low temperatures down to 100 °C and low log{$[H_2O]/[BiI_3]$} value down to 1. Therefore, it is thought that the O_2 should be the best oxygen source in the CVD process for BiOI or Bi_2O_3 from BiI_3. Actually, their experimental results suggest that the stability region of BiOI can also be stable at the temperature around 500 °C and log{$[O_2]/[BiI_3]$} value up to and 4.7, as depicted in Fig. 5.9.

5.4 Hydrothermal Preparation/Growth Process

For the hydrothermal growth of BiOX crystals, the bismuth salt such as $Bi(NO_3)_3$ or $BiCl_3$ was usually adopted as the staring materials. The BiOX nuclei could be formed in the presence of a halide source like halide salt.

Xiong et al. [20] developed a mannitol-assisted hydrothermal route for the preparation of two-dimensional BiOCl nanoplates with square crystal edge, using $Bi(NO_3)3$ and saturated NaCl solution as the starting reagent. As illustrated in Scheme 5.2, the layered structure of BiOCl tends to have a plate-growth in order to achieve a rapid growth along [001] direction with the help of mannitol, and in the meantime the hydrogen bonding between hydroxyl groups and its selective adsorption also help the formation of regular square-like nanoplates.

Deng et al. [21] proposed a hydrothermal route for the preparation of BiOX (X = Cl, Br) with two-dimensional structure with controllable particle size (both in-plane size and thickness) in the mixture of PVP and H_2O_2 aqueous solution. They used metal bismuth powder and NaXas as the chemical sources for bismuth and halide. An oxidation-reduction process between Bismuth metal and H_2O_2 takes place in the hydrothermal conditions. The chemical reaction involved in the formation bismuth oxyhalides is shown below:

$$2Bi + 3H_2O_2 + 2NaX \rightarrow 2BiOX + 2NaOH + 2H_2O.$$

Actually, the spherical polycrystalline nanoparticles of bismuth oxhydrochlorides (Bi(OH)Cl) with particle size ranging between 5 and 20 nm can be obtained before the hydrothermal reaction; however, the corresponding XRD pattern is not indexed in the literature or JCPDS card to date. As illustrated in Scheme 5.3, by increasing the hydrothermal temperature and concentrations of the Bi precursor, the particle size of BiOX can be increased from nanoscale to microscale and this leads to the side-by-side self-assembling from Bismuth oxhydrohalides nanoparticles to Bismuth oxyhalide nanoplates, and also the piling up of the two-dimensional Bismuth oxyhalide crystals.

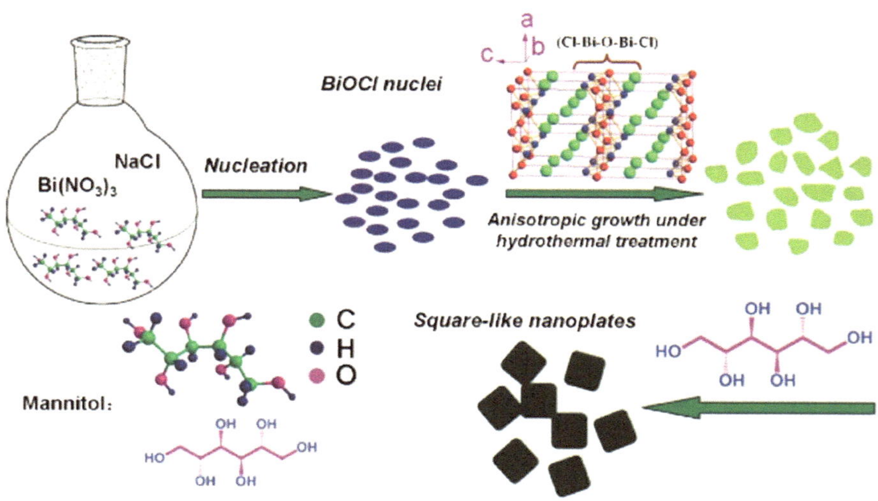

Scheme 5.2 Illustration of the possible formation mechanism of square-like BiOCl nanoplates. Reprinted with permission from Ref. [20]

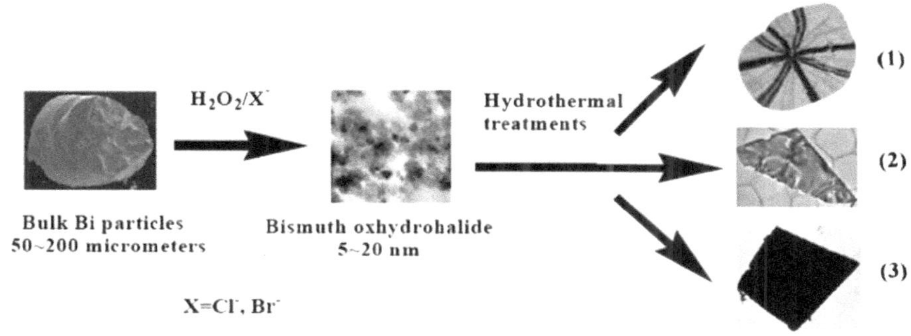

No.	BiOX Product	In-plane size	Thickness	Temperature	Concertration
1	nanoplates	200~500nm	10~30 nm	160 °C	0.1 mmol
2	nanosheets	2~50 μ m	40~80 nm	180 °C,	0.2 mmol
3	microsheets	10~100 μ m	200~1000 nm	200 °C	0.4 mmol

Scheme 5.3 Illustration of large-scale in-plane size and thickness controllable synthesis of single-crystalline BiOX (X = Cl, Br) Micro- and nanostructures (*Top*) and the experimental parameters and the description of the products described in the scheme (*Bottom*). Reprinted with permission from Ref. [21]

5.5 Solvothermal Preparation/Growth Process

5.5.1 Alcoholic-Mediated Process

To the best our knowledge, Zhu et al. [22] first reported the preparation of bismuth oxyhalide compound (BiOCl) by solvothermal method in 2002. In their study, $BiCl_3$ was used as the starting material to prepare polycrystalline and single-crystalline BiOCl in absolute ethanol through a solvothermal process at 120–15 °C. In this process the reaction temperature plays an important role in the crystal structure of BiOCl, where the evolution of BiOCl crystals from powder-like crystal to single-crystal were found as the reaction temperature was increased from 120 to 150 °C. Also the hydrolysis process also takes place due to the traces of water in the solvent of absolute ethanol.

Yuan et al. [23] adopted similar route for the preparation of BiOCl fibers, where activated carbon fibers were used as the template in the reaction, in addition to the reaction material $BiCl_3$ and anhydrous ethanol media. The lava-like rough surface and solid interior structure could be prepared through ethanol-mediated solvothermal process. Also it is interesting to note that when acetone is used in the place of ethanol for the solvothermal process, BiOCl fibers with completely different morphology were obtained. The morphology of BiOCl fibers shows the fiber wall with hollow interior consisting of numerous lamellas. The XRD patterns of BiOCl fibers prepared through the ethanol as medium display the complete characteristic diffraction peaks of BiOCl, whereas a few typical peaks of BiOCl are found in the sample prepared in acetone medium, implying the effect of solvent on the morphology and orientation of the crystal growth.

Ethylene glycol (EG)-mediated self-assembly process has been intensively investigated for the preparation of three-dimensional nanostructures, even though the reason for such an unusual morphology is not yet well understood [24, 25].

The chemical structure of EG is depicted in Fig. 5.10. For example, in the preparation of BiOX from $Bi(NO_3)_3$ with the EG mediation, the coordination reaction between EG and Bi^{3+} takes place during the solvothermal process and in this reaction bismuth alkoxides and HNO_3 are, respectively, the product and by-product. The bismuth alkoxides precipitate and become the nuclei for the growth of nanoparticles with centered distribution, and the presence of HNO_3 inhibits the further formation of bismuth alkoxides and BiOCl. In addition to the role of EG as a solvent, it can also function as a soft template in the solvothermal preparation, because of the hydrogen bonds between hydroxyl groups [26].

In the synthesis of BiOX three-dimensional nanostrcutures, the EG-mediated solvothermal process leads to Ostwald ripening and self-assembly process by the following steps: (i) BiOX crystalline nuclei formed in the presence of EG by solvothermal process, (ii) the crystalline nuclei of BiOX further grow up and freely aggregate to minimize their surface area by Ostwald ripening; and (iii) the crystalline nuclei of BiOX grow into two-dimensional nanosheets according to the inner crystallographic orientations.

The first report on the solvothermal preparation route for BiOCl came out in 2002 [22], and after a small pause a new spurt of enthusiasm on the solvothermal route started in 2008. Since then, successful preparation of BiOX (X = Cl, Br, I) with hierarchical structure in ethylene glycol-mediated solvothermal route, using Bi $(NO_3)_3$ and halide salt (KCl, NaBr, KI) as the staring materials, have been reported [27, 28].

Zhu et al. [29] synthesized three-dimensional BiOCl hierarchitectures through EG-mediated solvothermal process using urea as the hydrolytic agent, which provides a steady OH^- supply through its hydrolysis but without templates or surfactants. It is deduced that the coordination reaction between Bi^{3+} and EG is further supported by of hydrolysis of urea, which neutralize the HCl by-product. It should be noted that no precipitation was found without hydrolytic agent, urea in the

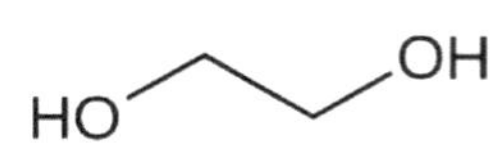

Fig. 5.10 The chemical structure of Ethylene glycol

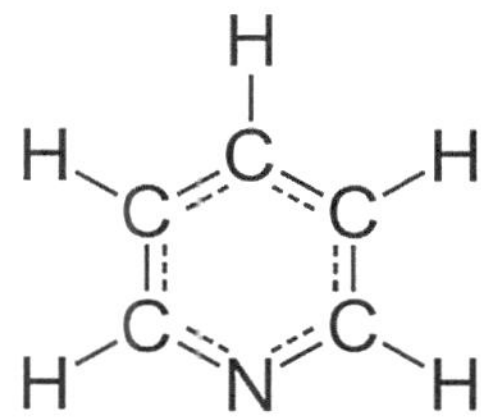

Fig. 5.11 The chemical structure of pyridine. Reprinted with permission from Ref. [30]

identical solvothermal conditions and also the crystal phase of the final product is strongly dependent on the molar ratio of urea/Bi^{3+}. It was noticed that BiOCl phase can be obtained with the optimum molar ratio of urea/Bi^{3+} of about 1–5 range. However, the crystal phase of α-Bi_2O_3 and $Bi_{12}O_{17}C_{12}$ were observed when the mole ratio of urea/Bi^{3+} increases from 5 to 10.

5.5.2 Others Solvothermal Systems

Song et al. [30] prepared hierarchical structured BiOCl via a pyridine-mediated solvothermal route, using $BiCl_3$ (without crystalline water) as the raw material. From their work, it was found that the traces of water present in pyridine and adsorbed water on $BiCl_3$ leads to the reaction so that the BiOCl is formed. Their results clearly indicate that the thickness of the BiOCl nanoplates, which assembles to hierarchical structure, depends on the volume ratio of H_2O/pyridine. The biggest thickness of the nanoplates can be obtained in pure pyridine solvent system and the thickness of the BiOCl nanoplates deceases as this volume ratio decreases. As shown in Fig. 5.11, a nitrogen atom in every pyridine molecule replaces one C–H group in the six-membered ring. Therefore, it is reasonable to deduce that a basic lone pair of electrons from the nitrogen atom leads to the competition with Cl^- in the formation process of BiOCl and thereby influencing the morphology of the BiOCl products.

5.6 Special Halide Source-Induced Preparation

5.6.1 Reactive Ionic Liquid

In addition to traditional halide-containing chemicals like sodium/potassium halide (KX) used as the halide source, since 2010, the ionic liquid (IL) was used for the preparation of BiOX compounds. This is known as ionothermal synthesis, which uses IL simultaneously as a solvent, a potential template and structure-directing agent in the growth of crystals, is similar to hydrothermal or solvothermal synthesis.

In effect, IL can function not only just as both solvent and template in the ionothermal synthesis, but also affect the chemical reaction process.

Ma et al. [31] found that the temperature plays a vital role in the size of BiOCl ultrathin nanoflakes (as thin as around 18 nm) formed through ionothermal synthesis route using $Bi(NO_3)_3$ and 1-hexadecyl-3-methylimidazolium chloride ([C_{16}Mim]Cl, as the raw materials as depicted in Fig. 5.12a), where IL of [C_{16}Mim]Cl functions as solvent, reactant and soft template in this reaction.

As illustrated in Scheme 5.4, the preferred sorption between the *ab* plane of BiOCl and $[C_{16}Mim]^+$ cations is quite convincing one, as the [BiOCl] layers are stacked by relatively weak Van der waals forces. The crystalline growth in the [001] direction in BiOCl is inhibited due to the selective sorption of $[C_{16}Mim]^+$ cations onto (001) crystal plane, which results in the formation of ultrathin nanoflakes.

Similarity, for the preparation of BiOBr, IL of 1-hexadecyl-3-methylimidazolium bromide ([C_{16}mim]Br, Fig. 5.12) is usually taken as the starting material. Cheng et al. [32] reported a facile one-pot approach to prepare uniform BiOBr hollow microspheres (HMSs) through solvothermal reaction in the mixture of 2-methoxyethanol and the IL of [C_{16}Mim]Br. Actually, emulsions are formed before the solvothermal reaction as the limited miscibility of [C_{16}Mim]Br and 2-methoxyethanol. As illustrated in Scheme 5.5, a microreaction chamber was formed by the microsized miniemulsions surrounded by 2-methoxyethanol containing Bi^{3+}. BiOBr nuclei could form as the reaction between Bi^{3+} and IL of [C_{16}Mim]Br at the interface of miniemulsions, followed by the preferential formation of two-dimensional nanosheets of BiOBr crystal. Finally a further growth of BiOBr nanosheets at the interface of miniemulsion, results in the formation of microsphere shells.

Xia et al. [33] developed a one-pot EG-assisted solvothermal process in the presence of reactive ionic liquid 1-hexadecyl-3-methylimidazolium bromide

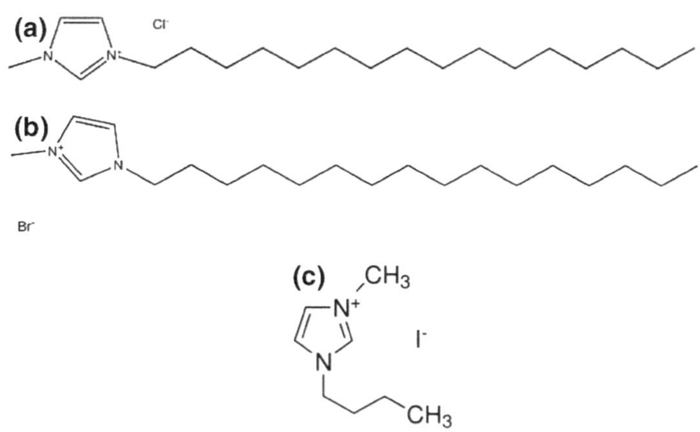

Fig. 5.12 Chemical structures of IL [C_{16}Mim]Cl (**a**), [C_{16}Mim]Br (**b**) and [Bmim]I (**c**). Reprinted with permission from Ref. [31]

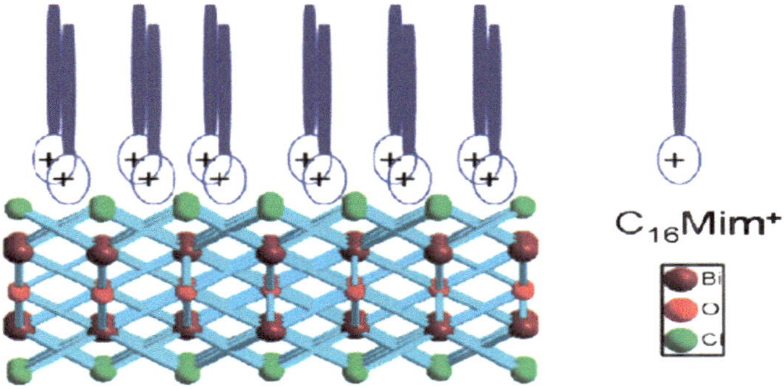

Scheme 5.4 Schematic illustration of the interaction of BiOCl crystal planes and the head parts of [C_{16}Mim]Cl. Reprinted with permission from Ref. [31]

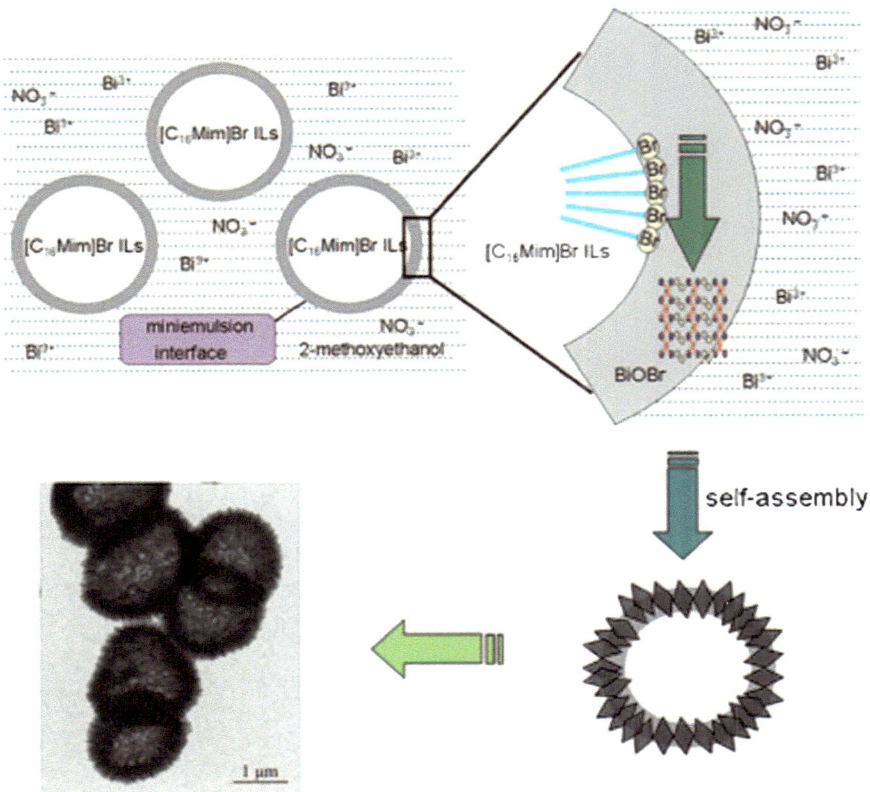

Scheme 5.5 A plausible schematic formation process of the BiOBr HMSs at the miniemulsion interface. Reprinted with permission from Ref. [32]

([C_{16}mim]Br, as depicted in Fig. 5.12) for the synthesis of BiOBr with the structure of uniform flower-like hollow microsphere. The ionic liquid micelles provide a soft template for the formation of BiOBr nuclei and the possible formation mechanism is illustrated in Scheme 5.6.

Also, Xia et al. [34] prepared BiOI hollow microspheres through a EG-mediated solvothermal route, using $Bi(NO_3)_3$ and IL of [Bmim]I (1-butyl-3-methylimidazolium iodine, as depicted in Fig. 5.12c) as the raw materials. It was deduced that the IL of [Bmim]I plays a significant role as a soft template in the formation of hollow structure possibly through the following steps (as illustrated in Scheme 5.7), (i) nuclei of BiOI formed on the surface of IL micelles, (ii) BiOI

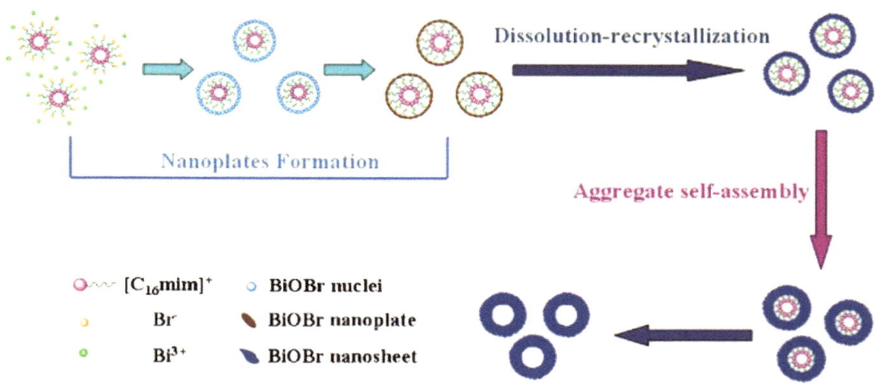

Scheme 5.6 Schematic illustration of proposed formation mechanism of BiOBr hollow structures. Reprinted with permission from Ref. [33]

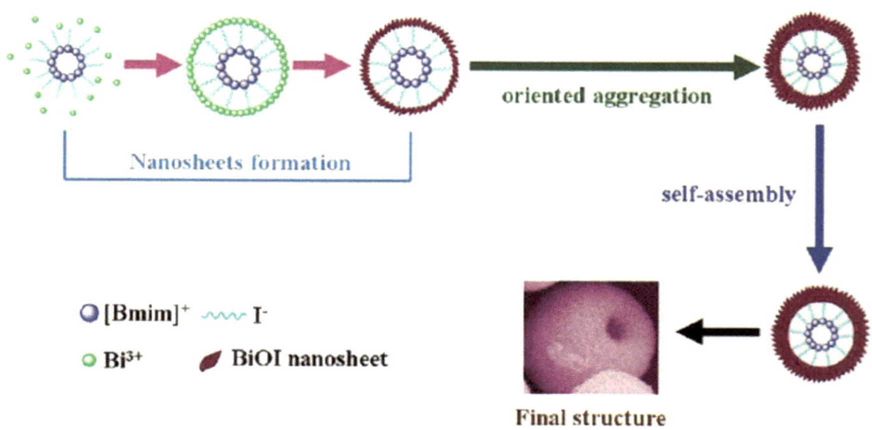

Scheme 5.7 Schematic illustration of proposed formation mechanism of BiOI hollow structures. Reprinted with permission from Ref. [34]

nuclei grew and preferentially transformed into two-dimensional nanosheets on the surface of IL micelles and is further aggregated into flower-like three-dimensional structure driven by Ostwald ripening, (iii) flower-like BiOI structures further transformed into hollow structure via a self-assembly process, and (iv) the increased pressure difference between the inner and outer of hollow structures after the reaction complement pushed the IL out from the unsubstantial point of BiOI crystal, resulting the formation of the hole in the product. On the other hand, it was found that the BiOI grew much faster in the same reaction time than BiOBr which prepared by their group before (using IL of $[C_{16}mim]Br$), possibly owing to the shorter alkylcarbon chain in [Bmim]I (i.e. $[C_4mim]I$), which could facilitate the crystal growth.

Others interesting investigation by this group [33–35] shows that when combining polyvinylpyrrolidone (PVP) with IL of $[C_{16}mim]Br$ or [Bmim]I in the reaction system, the solid and porous nanospheres or microspheres of BiOBr and BiOI can, respectively, be obtained, rather than a hollow structure. However, the possible mechanism is still not clear.

5.6.2 *Reactive Surfactant*

In recent years, some halide-containing cationic surfactants like Cetyltrimethylammonium chloride or bromide (CTAC or CTAC, as depicted Fig. 5.13) are used as halide source and soft template for the preparation of Bismuth oxyhalide compounds. Such synthesis is normally carried out in the hydrothermal or solvothermal conditions.

In early 2009, Shang et al. [36] proposed a hydrothermal reaction-based preparation of BiOBr with lamellar structure in aqueous phase using CTAB as Br source and template, through the following chemical reactions:

$$Bi^{3+} + C_{16}H_{33}(CH_3)_3N{-}Br \rightarrow C_{16}H_{33}(CH_3)_3N{-}Br{-}Bi^{3+}$$

$$C_{16}H_{33}(CH_3)_3N{-}Br{-}Bi^{3+} + H_2O \rightarrow BiOBr + C_{16}H_{33}(CH_3)_3N^{+} + 2H^{+}$$

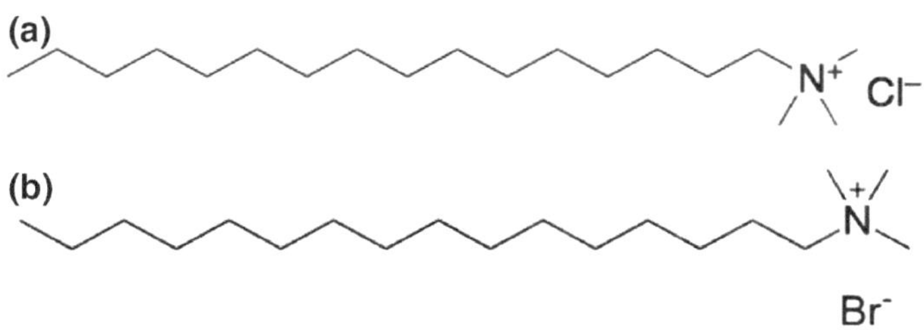

Fig. 5.13 Chemical structure of CTAC (**a**) and CTAB (**b**)

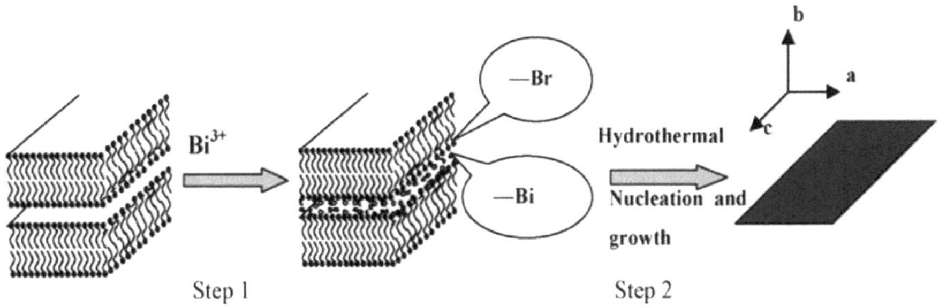

Scheme 5.8 The formation mechanism of the lamellar BiOBr. Reprinted with permission from Ref. [36]

The yellow suspension, which is considered to be the complex of Bi^{3+} and CTAB (i.e., $C_{16}H_{33}(CH_3)_3N–Br–Bi^{3+}$), was formed immediately after mixing acid Bi^{3+} and CTAB in aqueous solution, and this is followed by the nucleation process of BiOBr and further growth and shaping into the lamellar structure under the assistance of CTAB as the soft template (as illustrated in Scheme 5.8).

It is interesting to notice that Xiao et al. [37] prepared BiOBr with hierarchical structures via a modified hydrothermal process that is similar to the reaction system reported by Shang et al. [36], but inducing NaOH in the aqueous phase. Their results further confirm that only BiOBr nanoflakes were obtained without NaOH in the reaction system, indicating the influence of NaOH amount on the morphology of the product, which is still not clear. A possible reaction pathway is proposed as below:

$$Bi(NO_3)_3 + H_2O \rightarrow BiONO_3 + 2H^+ + 2NO_3^-$$
$$BiONO_3 + Br-(CTAB) \rightarrow BiOBr + NO_3^-$$

It is considered that, first, the $BiONO_3$ is formed due to the hydrolysis of Bi $(NO_3)_3$ under hydrothermal conditions, followed by a ion exchange process between NO_3^- and Br^- (from CTAB). However, this reaction process proposed by Xiao et al. is different from the one proposed by Shang et al. [36], where the formation of complex between Bi^{3+} and CTAB cannot be ignored.

Besides the effect of NaOH on the morphology of the BiOBr product, the nature of solvent can also substantially influence the morphology of the product. It is worth noticing that when EG is used as a solvent in the place of water, a completely different morphology can be obtained. A few reports [38–41] have already shown that the BiOBr with hierarchical structures can be prepared through an alcohol compounds (such as EG or ethanol, etc.) mediated solvothermal process, using $Bi(NO_3)_3$ and CTAB as the raw materials which are almost identical to the reference [36].

Chen et al. [40] proposed a stepwise formation process of hierarchical BiOBr, which is so-called the "stepwise blossoming." As illustrated in Scheme 5.9, it is considered that the following steps can take place during the reaction: (i) the BiOBr

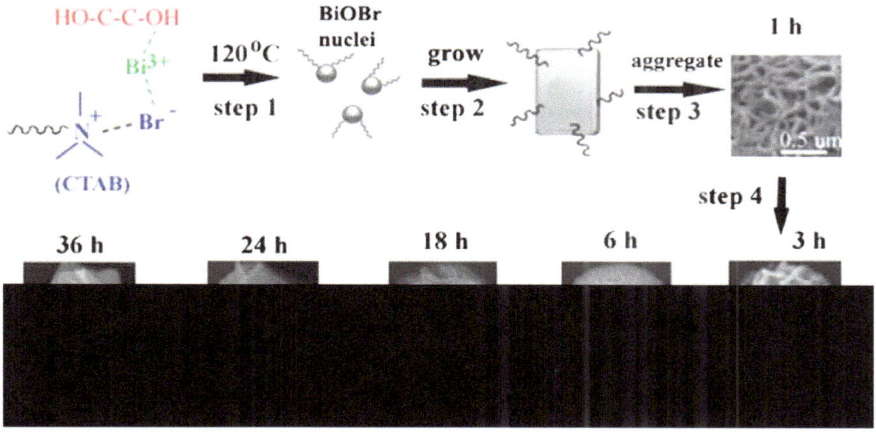

Scheme 5.9 The formation process of stepwise formation of BiOBr nanoplate-assembled microflowers. Reprinted with permission from Ref. [40]

nuclei is formed under solvothermal conditions due to the electrostatic forces between the Bi^{3+} cations and Br^- (from CTAB) and partial negative charge of oxygen atoms in EG; (ii) the crystal nuclei of BiOBr grow into nanoplates anisotropically due to the intrinsic tetragonal layer-structure as well as due to the selective adsorption of CTAB on the crystal surfaces; (iii) individual building blocks assemble and transform into three-dimensional nanostructures through Ostwald ripening; and (iv) the three-dimensional nanostructures fully grow into a well-crystallized nanoplate-assembled microflowers.

In the solvothermal system of Bi^{3+}-EG-CTAB, CTAB is also found to have great effect on the size control of the BiOBr product as a soft template. As reported by Chen et al., it was found that BiOBr product with non-uniform particle-size distribution can be obtained if CTAB is replaced with traditional NaBr salt as the bromide source, possibly due to the selective adsorption of CTAB molecules on certain crystal planes which prevents the arbitrary growth of the BiOBr nuclei.

Using similar method, Liu et al. [42] and Chen et al. [43] prepared nano-BiOCl microsphere through an EG-mediated solvothermal route, using $Bi(NO_3)_3$ and CTAC as the bismuth and chloride sources, respectively, and the details of the formation mechanism were investigated and discussed in their papers, which can be applied to explain the mechanism of hierarchical BiOBr formation.

Various coordinated compounds formed between Bi^{3+} and EG molecules (such as $Bi(EG)_2$) were analyzed using LC-MS depicted in Fig. 5.14. Also it is reasonable to expect that the formation of coordinated compound between Bi^{3+} and trace water in EG solvent. The formation of such complexes results in the drastic decrease of the pH value after adding $Bi(NO_3)_3$ into EG solvent. This process could be described as follows:

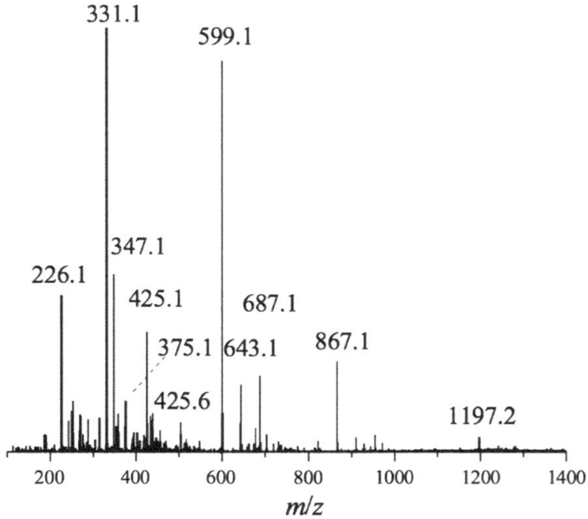

Fig. 5.14 Mass spectrum of the bismuth glycol solution. Reprinted with permission from Ref. [42]

$$xBi^{3+} + zH_2O + yEG \rightarrow Bi_x(EG)_y(OH)_z + zH^+$$
$$Bi_x(EG)_y(OH)_z \rightarrow Bi_x(i{-}EG)_y(OH)_z + yH^+$$

In this system, the interaction between Bi^{3+} and OH^- or EG molecules becomes stronger with the increase of pH value by accelerating the dissolution of Bi^{3+} and the formation of BiOCl, as shown below:

$$Bi_x(i{-}EG)_y(OH)_z + zOH^- + Cl^- \rightarrow xBiOCl + yEG + zH_2O.$$

It was found that white precipitation was formed when pH was increased from 0.38 (by adding $Bi(NO_3)_3$ into EG solvent) to 4.5 (by adding KOH aqueous solution), and this indicates that OH^- gradually take part in the hydrolysis of Bi^{3+} by replacing EG ligand. Therefore, it is clear that the conventional BiOCl flakes obtained under higher pH are originated from the spontaneous growth of BiOCl nuclei. On the contrary, under low pH condition, Bi^{3+} cannot exist as complex form until undergoing a hydrothermal process at 160 °C, where the dehydration catalyzed by Bi^{3+} promotes the hydrolysis of Bi^{3+} and formation of BiOCl. The hierarchical BiOCl nanostructures can be obtained by the slow growth of Bi^{3+} into BiOCl on the crystalline nuclei of BiOCl due to the coordinated interaction between Bi^{3+} and EG molecules (or its condensates).

5.7 Other Preparation Methods

Besides the preparation methods mentioned above, some others preparation techniques such as electrospinning [44], ultrasound [45], microwave [46], reverse microemulsion [47], etc., have been studied in order to obtain the desired morphology, which make the preparation of such series bismuth oxide halide compounds more fruitful.

References

1. Keller E, Krämer V (2005) A strong deviation from Vegard's rule: X-Ray powder investigations of the three quasi-binary phase systems BiOX–BiOY (X, Y = Cl, Br, I). Zeitschrift für Naturforschung B 60:1255–1263
2. Zheng GQ, Tang MT (2000) Phases of residue in distillation of BiCl3–HCl–H2O system. Chin J Nonferrous Met 10:250–252
3. Wang Y, Peng W, Chai L (2004) Thermodynamic equilibrium of bismuth hydrometallurgy in chloride and nitrate solutions. J Cent South Univ Technol 11:410–413
4. Wosylus A, Hoffmann S, Schmidt M et al (2010) In-situ study of the solid-gas reaction of $BiCl_3$ to BiOCl via the intermediate hydrate $BiCl_3 \cdot H_2O$. Eur J Inorg Chem 2010:1469–1471
5. Peng HL, Chan CK, Meister S et al (2009) Shape evolution of layer-structured bismuth oxychloride nanostructures via low-temperature chemical vapor transport. Chem Mater 21:247–252
6. Novokreshchenova MN, Yukhin Y, Bokhonov BB (2005) Highly pure bismuth (III) oxochloride synthesis. Chem Sustain Deve 13:563–568
7. Ye LQ, Zan L, Tian LH et al (2011) The 001 facets-dependent high photoactivity of BiOCl nanosheets. Chem Commun 47:6951–6953
8. Lee KS, Myung ST, Amine K et al (2009) Dual functioned BiOF-coated $Li[Li_{0.1}Al_{0.05}Mn_{1.85}]O_4$ for lithium batteries. J Mater Chem 19:1995–2005
9. Kang HB, Myung ST, Amine K et al (2010) Improved electrochemical properties of BiOF-coated 5V spinel $Li[Ni_{0.5}Mn_{1.5}]O_4$ for rechargeable lithium batteries. J Power Sources 195:2023–2028
10. Su WY, Wang J, Huang YX et al (2010) Synthesis and catalytic performances of a novel photocatalyst BiOF. Scripta Mater 62:345–348
11. Wang M, Huang QL, Chen XT et al (2007) Room temperature synthesis of bismuth oxyfluoride nanosheets and nanorods. Mater Lett 61:4666–4669
12. Bervas M, Yakshinskiy B, Klein LC et al (2006) Soft-chemistry synthesis and characterization of bismuth oxyfluorides and ammonium bismuth fluorides. J Am Ceram Soc 89:645–651
13. Ye LQ, Tian LH, Peng TY et al (2011) Synthesis of highly symmetrical BiOI single-crystal nanosheets and their 001 facet-dependent photoactivity. J Mater Chem 21:12479–12484
14. Yu CL, Fan CF, Yu JC et al (2011) Preparation of bismuth oxyiodides and oxides and their photooxidation characteristic under visible/UV light irradiation. Mater Res Bull 46:140–146
15. Ueda W, Sakyu F, Isozaki T et al (1991) Catalytic oxidative dimerization of methane to form C_2-compounds over Arppe's phase oxychlorides of Bi, La and Sm. Catal Lett 10:83–90
16. Silvestri VJ, Sedgwick TO, Landermann JB (1973) Vapor growth of $Bi_{12}GeO_{20}$, γ-Bi_2O_3 and BiOCl. J Cryst Growth 20:165–168
17. Keramidas KG, Voutsas GP, Rentzeperis PI (1993) The crystal structure of BiOCl. Z Kristallogr 205:35–40
18. Ganesha R, Arivuoli D, Ramasamy P (1993) Growth of some group V–VI–VII compounds from the vapour. J Cryst Growth 128:1081–1085
19. Schuisky M, Hårsta A (1998) Chemical vapor deposition stability diagram for the BiI_3–O_2 system. Electrochem Soc 145:4234–4239
20. Xiong J, Cheng G, Li G et al (2011) Well-crystallized square-like 2D BiOCl nanoplates: mannitol-assisted hydrothermal synthesis and improved visible-light-driven photocatalytic performance. RSC Adv 1:1542–1553

21. Deng ZT, Chen D, Peng B et al (2008) From bulk metal Bi to two-dimensional well-crystallized BiOX (X = Cl, Br) micro- and nanostructures: synthesis and characterization. Cryst Growth Des 8:2995–3003
22. Zhu LY, Xie Y, Zheng XW et al (2002) Growth of compound Bi^{III}–VI^{A}–VII^{A} crystals with special morphologies under mild conditions. Inorg Chem 41:4560–4566
23. Yuan RS, Lin C, Wu BC et al (2009) Synthesis of SnO_2, Fe_2O_3, and BiOCl fibers from inorganic salts by a templating route. Eur J Inorg Chem 2009:3537–3540
24. Zhong LS, Hu JS, Liang HP et al (2006) Self-assembled 3D flowerlike iron oxide nanostructures and their application in water treatment. Adv Mater 18:2426–2431
25. Cao AM, Hu JS, Liang HP et al (2005) Self-assembled Vanadium Pentoxide (V_2O_5) hollow microspheres from nanorods and their application in lithium-ion batteries. Angew Chem Int Ed 44:4391–4395
26. Wang YL, Jiang XC, Xia YN (2003) A solution-phase, precursor route to polycrystalline SnO_2 nanowires that can be used for gas sensing under ambient condition. J Am Chem Soc 125:16176–16177
27. Zhang X, Ai ZH, Jia FL et al (2008) Generalized one-pot synthesis, characterization, and photocatalytic activity of hierarchical BiOX (X = Cl, Br, I) nanoplate microspheres. J Phys Chem C 112:747–753
28. Zhang J, Shi FJ, Lin J et al (2008) Self-assembled 3-D architectures of BiOBr as a visible light-driven photocatalyst. Chem Mater 20:2937–2941
29. Zhu LP, Liao GH, Bing NC et al (2010) Self-assembled 3D BiOCl hierarchitectures: tunable synthesis and characterization. CrystEngComm 12:3791–3796
30. Song JM, Mao CJ, Niu HL et al (2010) Hierarchical structured bismuth oxychlorides: self-assembly from nanoplates to nanoflowers via a solvothermal route and their photocatalytic properties. CrystEngComm 12:3875–3881
31. Ma JM, Liu XD, Lian JB et al (2010) Ionothermal synthesis of BiOCl nanostructures via a long-chain ionic liquid precursor route. Cryst Growth Des 10:2522–2527
32. Cheng HF, Huang BB, Wang ZY et al (2011) One-pot miniemulsion-mediated route to BiOBr hollow microspheres with highly efficient photocatalytic activity. Chem Eur J 17:8039–8043
33. Xia JX, Yin S, Li HM et al (2011) Improved visible light photocatalytic activity of sphere-like BiOBr hollow and porous structures synthesized via a reactable ionic liquid. Dalton Trans 40:5249–5258
34. Xia JX, Yin S, Li HM et al (2011) Self-assembly and enhanced photocatalytic properties of BiOI hollow microspheres via a reactable ionic liquid. Langmuir 27:1200–1206
35. Xia JX, Yin S, Li HM et al (2011) Enhanced photocatalytic activity of bismuth oxyiodine (BiOI) porous microspheres synthesized via reactable ionic liquid-assisted solvothermal method. Colloids Surf A Physicochem Eng Aspects 387:23–28
36. Shang M, Wang WZ, Zhang L et al (2009) Preparation of BiOBr lamellar structure with high photocatalytic activity by CTAB as Br source and template. J Hazard Mater 167:803–809
37. Xiao PP, Zhu LL, Zhu YC et al (2011) Selective hydrothermal synthesis of BiOBr microflowers and Bi_2O_3 shuttles with concave surfaces. J Solid State Chem 184:1459–1464
38. Ai ZH, Ho WK, Lee SC et al (2009) Efficient photocatalytic removal of NO in indoor air with hierarchical bismuth oxybromide nanoplate microspheres under visible light. Environ Sci Technol 43:4143–4150
39. Xu J, Meng W, Zhang Y et al (2011) Photocatalytic degradation of tetrabromobisphenol A by mesoporous BiOBr: efficacy, products and pathway. Appl Catal B Environ 107:355–362
40. Chen YJ, Wen M, Wu QS (2011) Stepwise blossoming of BiOBr nanoplate-assembled microflowers and their visible-light photocatalytic activities. CrystEngComm 13:3035–3039
41. Feng YC, Li L, Li JW et al (2011) Synthesis of mesoporous BiOBr 3D microspheres and their photodecomposition for toluene. J Hazard Mater 192:538–544
42. Liu HQ, Gu XN, Chen F et al (2011) Preparation of nano BiOCl microsphere and its fabrication machanism. Chin J Catal 32:129–134
43. Chen F, Liu HQ, Bagwasi S et al (2010) Photocatalytic study of BiOCl for degradation of organic pollutants under UV irradiation. J Photochem Photobiol A Chem 215:76–80

44. Wang CH, Shao CL, Liu YC et al (2008) Photocatalytic properties BiOCl and Bi_2O_3 nanofibers prepared by electrospinning. Scripta Mater 59:332–335
45. Yu CL, Zhou WQ, Yu JC (2011) Rapid fabrication of BiOCl(Br) nanosheets with high photocatalytic performance via ultrasound irradiation. Chin J Inorg Chem 27:2033–2038
46. Zheng L, Cao XF, Chen XT et al (2011) BiOBr hierarchical microspheres: microwave-assisted solvothermal synthesis, strong adsorption and excellent photocatalytic properties. J Colloid Interface Sci 354:630–636
47. Henle J, Simon P, Frenzel A et al (2007) Nanosized BiOX (X = Cl, Br, I) particles synthesized in reverse microemulsions. Chem Mater 19:366–373

Chapter 6
Modified Bismuth Oxihalide Semiconductors

Abstract This chapter describes different variants of modified bismuth oxyhalides in order to enhance the functionality of the material. Different material modifications, like hybrid system with metal where the metal functions as a co-catalyst and also induces the surface Plasmon effect, and the hybrid system with carbon and other semiconducting materials that form a heterojunction are elaborately discussed.

Keywords Surface plasmon resonance · Heterojunction · Metal–BiOX hybrid system · Carbon–BiOX hybrid system · Semiconductor–BiOX hybrid system

6.1 Metal–BiOX Hybrid System

6.1.1 Metal as a Co-catalyst

The noble metal nanoparticles loaded on the surface of semiconductors can restrain the recombination of the photoinduced carriers, which could augment the interfacial charge transfer and facilitate the photocatalytic efficiency of the metal/semiconductor photocatalyst [1]. With irradiation of visible light, the electrons in the semiconductor transfer from the valence band to the conduction band, leaving a hole behind in the valence band. When the semiconductor is coupled with a noble metal, the electrons from the semiconductor transfer to noble metal particles and this is because the energy of the Fermi level of semiconductor is higher than that of noble metal. Thus, the photogenerated electrons accumulate on the noble metal, which serve as trap centers and facilitates the charge separation. Also these photogenerated electrons can transfer to surface-absorbed oxygen to generate activated $\cdot O^{2-}$ [2]. Finally, the activated $\cdot O^{2-}$ gives rise to $\cdot OH$ through a range of reaction with H^+, which is the photocatalytic reduction reactions process. On the other hand, holes accumulated on the conduction band of semiconductor could also react with H_2O to generate $\cdot OH$, which is the photocatalytic oxidation reactions process. The reductive and oxidative $\cdot OH$ species and the photogenerated holes can effectively decompose the organics.

M.A. Gondal et al., *Novel Bismuth-Oxyhalide-Based Materials and Their Applications*, Advanced Structured Materials 76,
DOI 10.1007/978-81-322-3739-6_6

The photocatalytic process taking place in the metal/semiconductor combination is mainly governed by three factors: Schottky contact, ohmic contact, and surface plasma resonance. The Schottky junction is different from hetrojunction as the former is formed between the metal and the semiconductor while the latter is formed between two or more semiconductors. The space charge layer of Schottky junction is located in the semiconductor and the photoinduced process is initiated only in the semiconductor. Normally when the light falls on the p–n junction, the electrons are injected from the p-type material into n-type material and the holes transfuse from n-type material to p-type material and after certain accumulation, pass to the surface by diffusion. However, when the light falls on the Schottky junction, the photogenerated electrons from the semiconductor are transferred to the metal without any accumulation as in Fig. 6.1. The function of Schottky junction is not only to effectively separate the photogenerated charge-carriers, but also to quickly move the separated photoelectrons, which effectively restrains photogenerated electron-hole recombination. Also this Schottky barrier facilitates the contribution of holes in photocatalytic reaction.

Work function is an important factor to be considered in the design of a Schottky junction. As we know at absolute zero, all the energy levels below Fermi level E_f are filled and all above E_f are empty in metal, while the energy level is higher than the E_f is all empty and the external energy is required for the electron to flow from the metal. The work function of the metal Wm can be defined as the difference between the metallic Fermi level E_f and the energy level E_0 of electrons in the vacuum state and it is the measure of the binding strength of the electrons to the metal. Like metal, the work function of the semiconductor is also defined as $W_S = E_0 - (E_F)_S$, and this work function is related to the doping concentration in the semiconductor. When the n-type semiconductor comes in contact with the metal and when $W_m > W_s$, and $(E_f)_s > (E_F)_m$, electrons flow from the semiconductor to metal resulting in the formation of the space charge region on the semiconductor surface. Hence, the direction of electric field vector point from inside to the surface resulting in the electron energy on the semiconductor surface being higher than the same within the body and this causes the energy band to bend upwards and to form

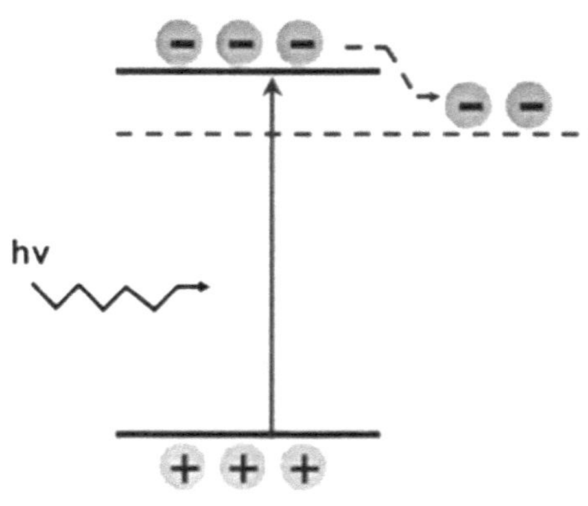

Fig. 6.1 Photogenerated charges transfer of the Schottky junction

a surface barrier. In the barrier, the space resistance is mainly formed by ionization benefactor and the electron concentration is much less than the body and this area of high resistance is known as the barrier layer. On the other hand, when $W_m < W_s$, in the n-type semiconductor, the electrons flow from metal to semiconductor forming a space charge region in the semiconductor surface; and in this case, electric field direction points from surface to the body causing the energy band bending down. In this case, the electron concentration of the semiconductor surface is much bigger than the body and this layer of high conductivity is known as the barrier layer. For the contact between p-type semiconductor and metal, the conditions for the barrier layer formation are just the opposite to that of n-type discussed above. When $W_m < W_s$, the band bends upwards, forming p-type anti-blocking layer; when $W_m > W_s$, the band bends down forming p-type barrier. The existence of the semiconductor surface state for the same kind of semiconductor contact with different metal can form the barriers of different heights.

As indicated in Fig. 6.2, for n-type barrier layer, positive bias (metal anode, semiconductor by cathode) make the electron flow from semiconductor to metal, and allow the forward current from metal to semiconductor to increase; negative bias have little impact on reverse current. For p-type barrier layer, negative bias (metal anode) make the electron flow from metal to semiconductor, and allow the forward current from semiconductor to metal to increase; positive bias also take little on forward current. The rectifier contact and the function of the heterojunction can be judged by measuring the properties of the prepared samples.

However, not all the metal–semiconductor junctions form a rectifying Schottky barrier; a metal–semiconductor junction that conducts current in both directions without rectification, perhaps due to its schottky barrier being too low, it is called an ohmic contact. Ohmic contact means, it does not produce significant additional impedance and not significantly change the balance of carrier concentration inside

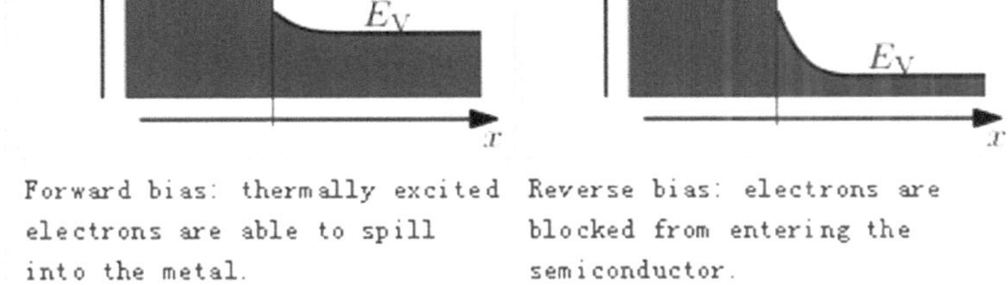

Fig. 6.2 Rectifier features of metal and semiconductor

the semiconductor. In terms of electricity, ideal ohmic contact resistance should be very small, compared with semiconductor sample. When a current flows through the junction, the pressure drop of ohmic contact voltage should be much smaller than the sample itself. This contact will not affect current–voltage properties of samples, that is to say the nature of the current–voltage is determined by the resistance characteristics of samples.

Both ohmic contact and Schottky barrier dependent on the Schottky barrier height, setting the threshold value of excess energy needs by electrons from the semiconductor to the metal. Ohmic contact is easy to accommodate electrons from both directions, but the barrier height must be less than some parts of the junction surface at least. Only with low barrier height everywhere and not reflecting electrons at the interfaces, it can perform like low resistance. The Schottky barrier height between a metal and semiconductor is naively predicted by the Schottky-Mott rule proportional to the difference of the metal–vacuum work function and the semiconductor–vacuum electron affinity. In practice, most metal–semiconductor interfaces do not follow this rule to the predicted degree. Instead, the chemical termination of the semiconductor crystal against a metal creates electron states within its band gap. The nature of these metal-induced gap states and their occupation by electrons tends to pin the center of the band gap to the Fermi level, an effect known as Fermi level pinning. Thus, the heights of the Schottky barriers in metal–semiconductor contacts often show little dependence on the value of the semiconductor or metal work functions, in stark contrast to the Schottky-Mott rule. Different semiconductors exhibit this Fermi level pinning to different degrees, but a technological consequence is that high quality (low resistance) ohmic contacts are usually difficult to form in important semiconductors such as silicon and gallium arsenide.

The Schottky-Mott rule is not entirely incorrect since, in practice, metals with high work functions form the best contacts to p-type semiconductors, while those with low work functions form the best contacts to n-type semiconductors. Unfortunately, experiments have shown that the predictive power of the model does not extend much beyond this statement. Under realistic conditions, contact metals may react with semiconductor surfaces to form a compound with new electronic properties. A contamination layer at the interface may effectively widen the barrier. The surface of the semiconductor may reconstruct leading to a new electronic state. The dependence of contact resistance on the details of the interfacial chemistry is what makes the reproducible fabrication of ohmic contacts such a manufacturing challenge.

6.1.2 Metal as a Surface Resonance Plasmon Probe

Surface plasmon resonance is a physical optical phenomenon. According to the theory of Drude, there are large numbers of free electrons in precious metals, such as Au, Ag, Cu, etc. When the absorbed light undergoes the total reflection at the interface between two phases, it can initiate those free electrons on the surface of the metal to produce the surface plasmon. If the angle of incidence and the wavelength of the light are of appropriate value, the resonance will occur and form the collective oscillation of electrons. Then the incident light is absorbed and reflected light energy falls sharply, and this resonance effect is shown in the reflectance spectral, [3] which is the surface plasmonic resonance effect.

As shown in Fig. 6.3, the wavelength of the electric field-induced free electron outside metal causes polarization, so that the polarity of spherical nanoparticles enhanced. Assume that positive charge of system is stabilized, and the free electron movement under the action of external electric field, therefore those negative charge in the inside of the metal particles under the electric field will be reset which leads to net charge on the surface of nanoparticles. The newly built electric field by net charge brings on the system a linear interior restoring force, thus produce the electronic dipole oscillation. This is the generation of surface plasmon resonance process.

Surface plasmon resonance depends on the size and shape of the metal nanoparticles. So we can adjust the size and shape of the metal nanoparticles to change the surface plasmon resonance effect. With the increase of metal particle size the absorption and scattering of light also will increase, and with the increase of absorption and scattering, scattering will gradually play a leading role and change the location and width of surface plasmon resonance. However, the regulation of metal particle size for the regulation of surface plasmon resonance is very limited. The shape of the particles is also having very important influence on surface plasmon resonance effect. In addition, the nanometer metal case can also adjust the surface plasmon resonance effect of noble metal particles [4]. With the increase of metal nanoparticles shell thickness, surface plasmon resonance absorption from visible light area redshift to near-infrared light area. This enhancement is mainly due to the coupling between the inner particle and the outer particle [5].

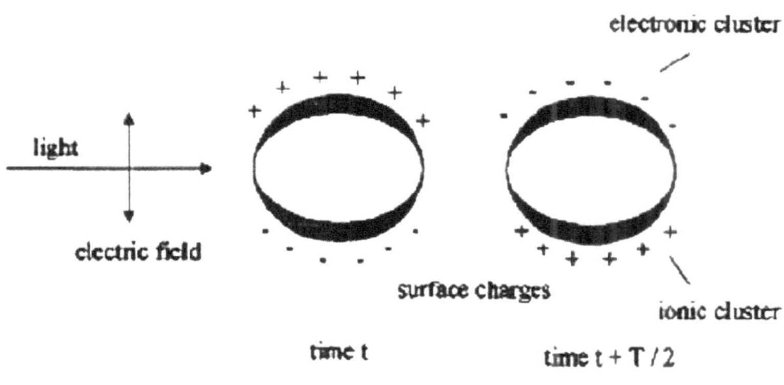

Fig. 6.3 Sketch of surface plasmon resonance effect on metal nanoparticles

It is noteworthy that the noble metal content in the semiconductor influences the efficiency of electron-hole separation and there exists an optimum level of metal content for obtaining the highest efficiency of any photocatalytic activity like photodegradation of dyes. When the noble metal content is below the optimum value, the noble metal particles on the semiconductor surface can act as electron-hole separation centers, and promote the electrons transfer to dioxygen or other electron acceptors. On contrast, when the noble metal content is above the optimum value, excessive negatively charged noble metal particles can catch holes and become the new recombination centers. Also the excessive noble metal content can lead to the aggregation of metal nanoparticles on the surface of semiconductor, which will give rise to the decrease of surface area and increase the scattering of visible light leading to the reduced photocatalytic efficiency of the photocatalytic reaction [6].

Liu et al. [7] synthesized Ag/BiOI nanoparticles which are active visible region. As shown in Fig. 6.4, when silver is coupled to BiOI and subject the material to visible light radiation, the electrons from BiOI transfer from the valence band to the conduction band leaving holes behind in the valence band. These electrons in the valence band transferred to silver particles and these accumulated electrons in the silver particles are subsequently transferred to surface-absorbed oxygen to generate activated O^{2-} and finally give rise to •OH. In the meantime, the holes accumulated in the valence band of BiOI react with H_2O to generate •OH. The reductive and oxidative •OH species and photogenerated holes effectively decompose the organic pollutants in the water.

In their study, it was found that the highest efficiency of the photodegradation of dyes was obtained when the concentration of silver in BiOI was 0.6%. This can be understood from the photoluminescence spectra depicted in Fig. 6.5, where the photoluminescence signal is lowest for 0.6% silver in BiOX. This indicates that the

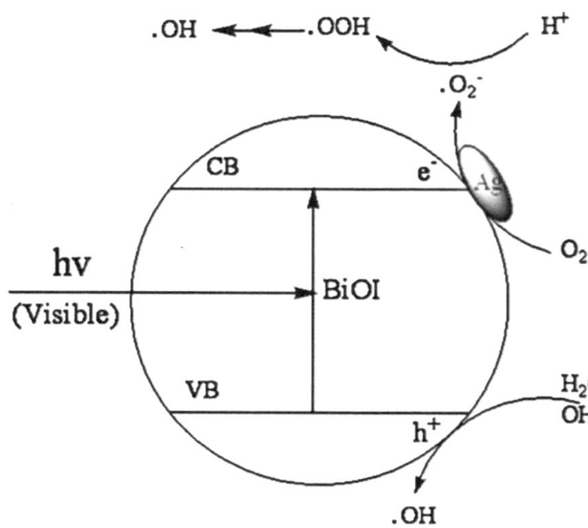

Fig. 6.4 Schematic diagram of the Ag/BiOI photocatalyst. Reprinted with permission from Ref. [7]

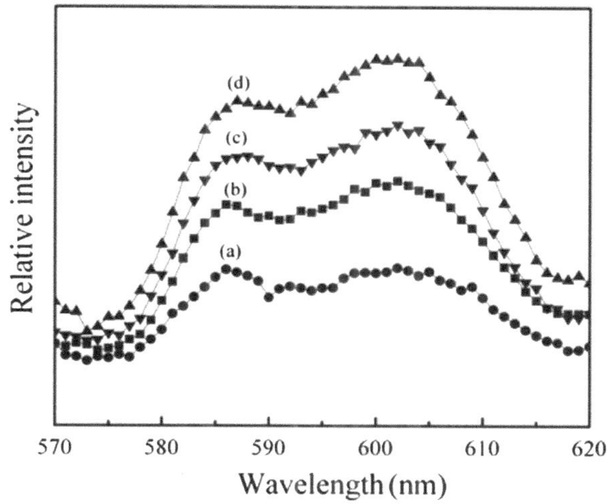

Fig. 6.5 PL spectra of the as-synthesized samples: **a** 0.6% Ag/BiOI, **b** 0.3% Ag/BiOI, **c** 0.9%Ag/BiOI and **d** pure BiOI. Reprinted with permission from Ref. [7]

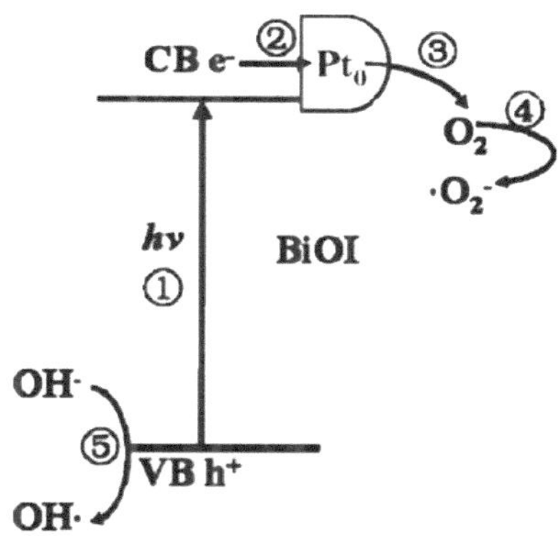

Fig. 6.6 Schematic diagram of the Pt/BiOI photocatalyst. Reprinted with permission from Ref. [8]

electron-hole recombination of self-trapped excitations at this optimum concentration is very low and all the charge-carriers are used for the degradation of dye through redox reactions. All the PL spectra in Fig. 6.5 were obtained at 390 nm wavelength laser excitation and the figure also has the PL spectra of other Ag concentrations in BiOX and also pure BiOI.

Yu et al. [8] prepared Pt/BiOI nanoplate catalyst and used this material for the photocatalytic degradation. Here the photoexcited electrons, the conduction band of BiOI, is transferred to the Pt atom; these electrons combine with surface-absorbed oxygen to generate activated $\bullet O^{2-}$ as the mechanism depicted in Fig. 6.6. In their study, it was observed that the highest efficiency of photodegradation of dye as found at an optimum Pt content is 0.2% in BiOI. Figure 6.7 shows the photoluminescence spectra of Pt/BiOX for different concentrations of Pt in BiOx, when excited with 320 nm laser wavelength. It is clear from Fig. 6.7 that the emission

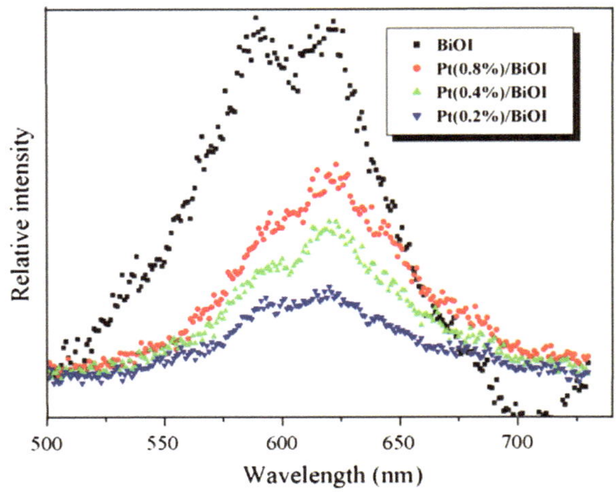

Fig. 6.7 Photoluminescence (PL) spectra of the samples. Reprinted with permission from Ref. [8]

intensity for Pt/BiOI is less compared to pure BiOI and also the emission intensities of the spectra vary for the different concentrations Pt contents in BiOI.

6.2 Carbon–BiOX Hybrid System

When BiOX is used as a photocatalyst, the electron recombination should be minimized in order to avail all the charge-carriers for the reaction and in order to inhibit the recombination of electron-hole pairs on a semiconductor, carbon-based materials like graphene, carbon nanotubes, and Fullerene are usually introduced. Graphene, a two-dimensional honeycomb structure, closely packed into a single layer of sp^2 ‾bonded carbon atoms, possess a high thermal conductivity (5000 W m^{-1} K^{-1}), excellent electron mobility (200,000 cm^2 V^{-1} s^{-1}), and high specific surface area (2600 m^2 g^{-1}). Nanomaterials can be dispersed onto graphene sheets because of the two-dimensional platform structure of graphene and it can also effectively retard the recombination of the electron-hole pairs in the graphene/BiOX composite due to the excellent electronic conductivity of graphene. Like in the case of metals discussed above, when graphene is combined with BiOBr [9], electrons flow from BiOBr to the graphene to align the Fermi energy levels at the interface of BiOBr and graphene. It would form a Schottky barrier at the interface of BiOBr and graphene. This Schottky barrier can catch electrons from BiOBr to graphene and inhibit their backflowing of electrons and as a result leading to effectively inhibition of electron-hole pair's recombination. As indicated in Fig. 6.8, electrons could be transferred from the VB to the CB under visible light irradiation, leaving holes at the VB of BiOBr. And then the photogenerated electrons transfer from the CB of BiOBr to graphene because of the strong chemical bonding of BiOBr and graphene. Due to the existence of Schottky barrier, the transportation of photoelectrons from graphene

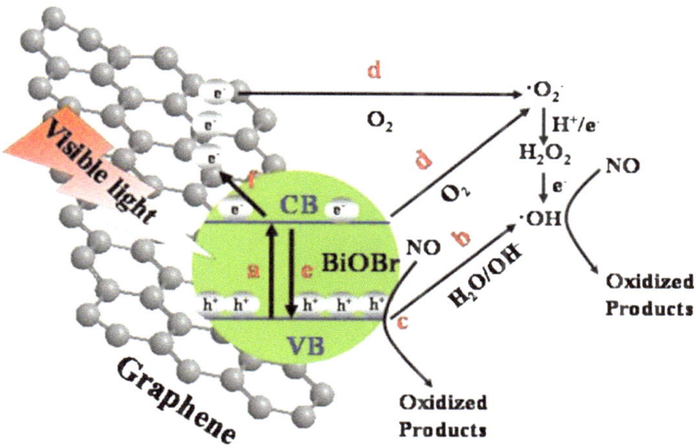

Fig. 6.8 Schematic illustration of the visible light photocatalytic enhancement of graphene/BiOBr composite. Reprinted with permission from Ref. [9]

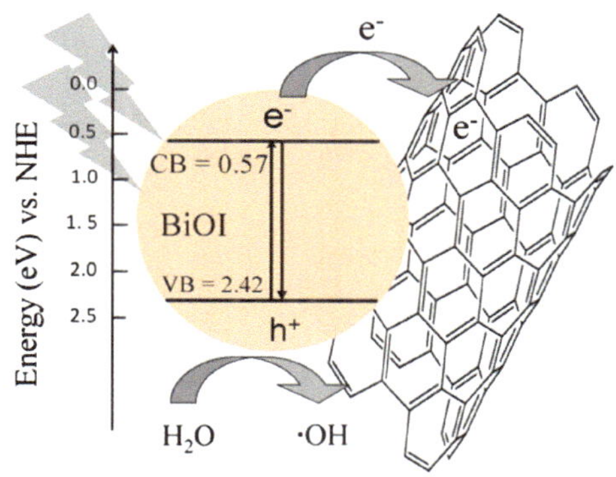

Fig. 6.9 Photoinduced electron transfer between an excited BiOI particle and MWCNT. Reprinted with permission from Ref. [10]

to the CB of BiOBr could be prevented, which improves the charge separation. These photogenerated electrons are transferred to surface-absorbed oxygen to generate activated $\cdot O^{2-}$. Finally, the activated $\cdot O^{2-}$ give rise to $\cdot OH$ through a series of reaction with H^+/e^-. Meanwhile, holes accumulated at the CB of BiOBr can also react with H_2O to generate $\cdot OH$. Generated $\cdot OH$ and photogenerated holes form could effectively oxidized the organics.

Carbon nanotubes, with a hollow geometry, possess a large specific surface areas and excellent charge transfer property. When carbon nanotubes are combined with BiOI [10], the role of carbon nanotubes in carbon nanotubes/BiOI is to capture and transport electrons, and thereby promoting the charge separation. Hence, the carbon nanotubes-doped BiOI is considered to be excellent composites with the outstanding synergetic effect of absorption and degradation for the organics. As indicated in Fig. 6.9, electrons are excited from the VB to the CB under visible light irradiation, leaving holes at the VB of BiOI. Subsequently, the photogenerated electrons transfer from the CB of BiOI to carbon nanotubes because of the strong chemical bonding of BiOI and carbon nanotubes, which inhibit the recombination of electron-hole pairs. Meanwhile, holes accumulated at the CB of BiOI could react with H_2O to generate •OH to effectively oxidize the organic pollutants.

Fullerene can lead to a rapid photoinduced charge separation and thereby reduce the charge recombination and this is due to the delocalized conjugated structures. Also the combination of C60 and nanoparticles promote the photoconversion of solar energy. When C60 is combined with Bi_2WO_6 [11], C60 plays a major role in C60/Bi_2WO_6 as an electron shuttle, which can result in the effective transfer of photoelectrons. As indicated in Fig. 6.10, with visible light irradiation electrons are excited from the VB to the CB, leaving holes at the VB of Bi_2WO_6. Subsequently, the photogenerated electrons transfer from the CB of Bi_2WO_6 to C60 because of the strong chemical bonding of Bi_2WO_6 and C60, which inhibit the recombination of electron–hole pairs. These photogenerated electrons are transferred to surface-absorbed oxygen to give rise to activated $\cdot O^{2-}$. Finally, the activated $\cdot O^{2-}$ produces •OH through a series of reaction with H^+. The photocatalytic reductive species •OH and photogenerated holes are pivotal in oxidizing the organic pollutants.

Gao et al. [12] synthesized graphene/BiOCl nanocomposites. As indicated in Fig. 6.11, with UV light irradiation, electrons could be transferred from the VB to the CB, leaving holes at the VB of BiOCl. And then the photogenerated electrons transfer from the CB of BiOCl to graphene, which improves the charge separation. These photogenerated electrons could be transferred to surface-absorbed oxygen to generate activated $\cdot O^{2-}$. Finally, the activated $\cdot O^{2-}$ give rise to •OH through a series of reaction with H^+/e^-. Meanwhile, holes accumulated at the CB of BiOCl could also react with H_2O to generate •OH. Generated •OH and photogenerated holes form could effectively oxidized the organics.

Tu et al. [13] synthesized BiOBr/graphene nanocomposites and observed that his material showed a substantial increase in the photocatalytic activity due to the charge separation accomplished through the mechanism described in the Fig. 6.12 and also as explained before.

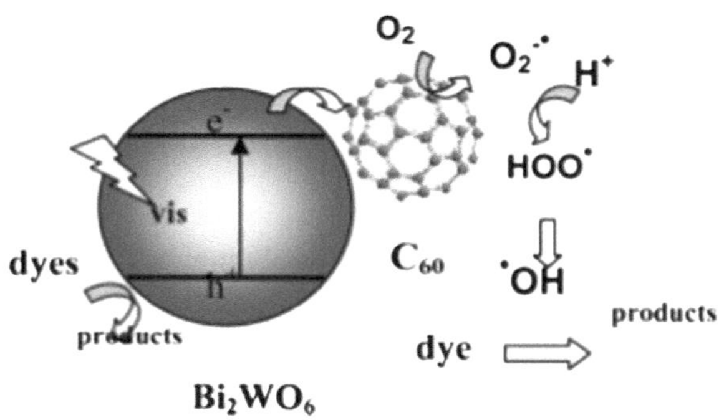

Fig. 6.10 Photoinduced electron transfer between an excited Bi_2WO_6 particle and C60. Reprinted with permission from Ref. [11]

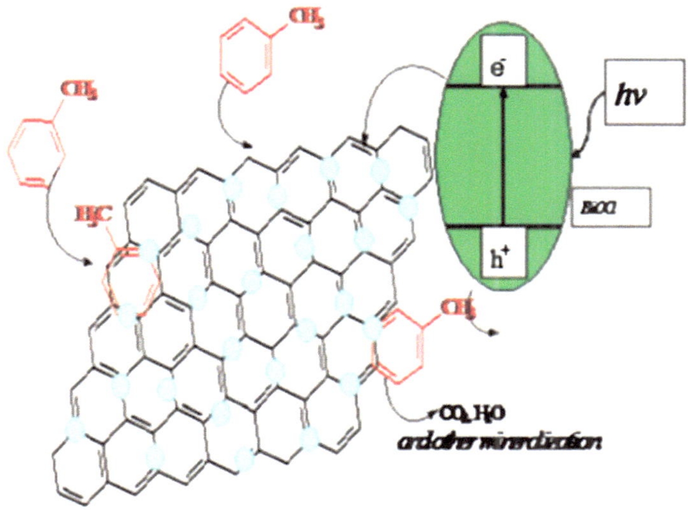

Fig. 6.11 Schematic illustration of the visible light photocatalytic enhancement of graphene/BiOCl composite. Reprinted with permission from Ref. [12]

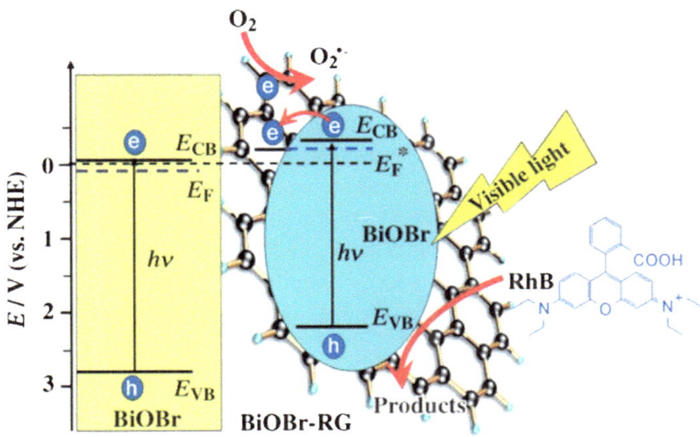

Fig. 6.12 Schematic structure of BiOBr–RG and photodegradation process of RhB over BiOBr–RG. Reprinted with permission from Ref. [13]

6.3 Semiconductor-BiOX Hybrid System

As explained in the beginning of this chapter, a heterojunctions are formed, when two semiconductors forms a composite and according to their respective band positions Semiconductor heterojunctions can be mainly divided into three types: straddling gap (type I), staggered gap (type II) and broken gap (type III), as

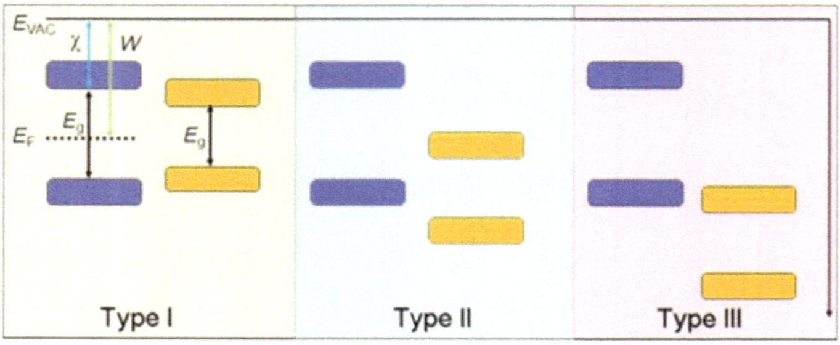

Fig. 6.13 Schematic diagram of the three types of semiconductor heterostructures

presented in Fig. 6.13. For any photocatalytic applications, staggered gap (type II) is mostly preferred and is known as A-type heterojunction.

In the case of A-type hertojunctions, sensitizers with narrow band-gap energies are coupled onto the surface of semiconductor to get a visible light-driven photocatalysts. When the visible light falls on these materials, the electrons in the VB of the sensitizer are excited to its CB, leaving the photogenerated positive carriers on VB of the sensitizer. Since the CB of the sensitizer is higher than that of semiconductor, the photogenerated electrons in the sensitizer transfer to the CB of semiconductor. However, on the other hand, the photogenerated positive carriers in the sensitizer cannot transfer to the VB of semiconductor, as its VB is positioned higher than that of semiconductor. So, in A-type heterojunction structure, the transferred electrons to the CB of semiconductor can be used to photocatalytic reduction reactions, but photocatalytic oxidation reactions cannot occur because the holes are not generated in the VB of semiconductor. Based on the discussions, The A-type heterojunction structure also can be used for the decomposition of organics, but complete mineralization of organic pollutant is difficult to implement.

When the VB level of the sensitizer is lower than that of semiconductor heterojunction structure, thus formed is essentially different from A-type heterojunction structure and is now known as B-type heterojunction as indicated in Fig. 6.14.

In this hybrid system [14], by absorbing visible light, the electrons in the VB of the sensitizer are excited to its CB, leaving the photogenerated holes on VB, and then the electrons in the CB of semiconductor can flow down to that of sensitizer due to its CB is positioned higher than that of sensitizer, finally, positive carriers are generated in the VB of semiconductor. Therefore, photocatalytic oxidation reactions can occur on VB of semiconductor, as a result the B-type heterojunction structure can induce complete mineralization of organic pollutant.

In addition to the above two heterojunction photocatalysts, another important category of semiconductor composites is p–n junction photocatalysts. [15]. As shown in the Fig. 6.15, the p-type semiconductor has both elections and holes as charge-carriers in which the number of the hole is greater than electrons. But the p-type semiconductor has fixed negative charge whose number is equal to the extra

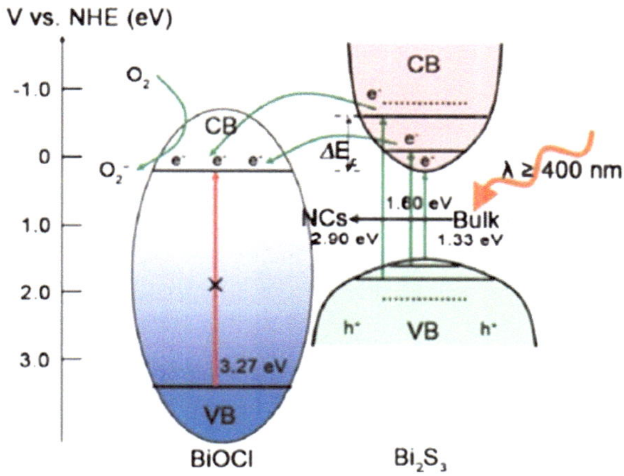

Fig. 6.14 Schematic diagram of the B-type heterojunction

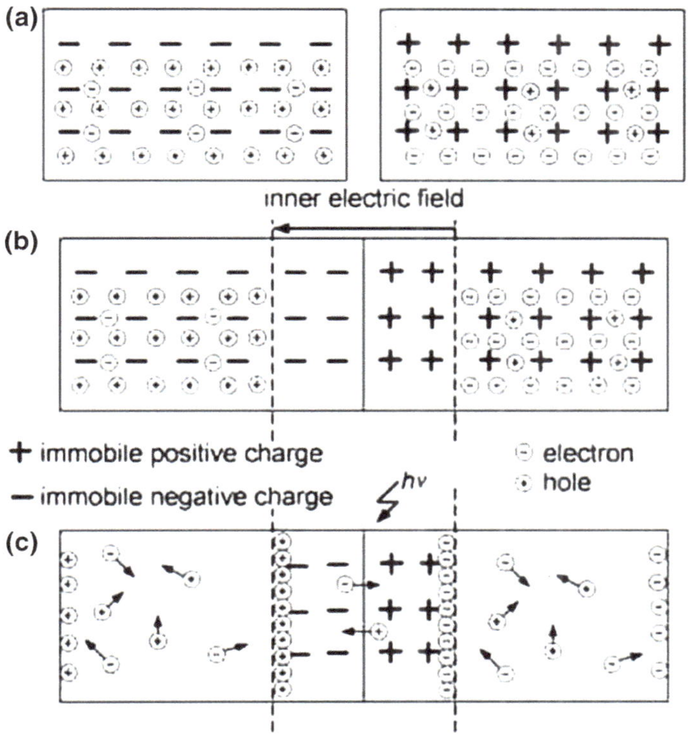

Fig. 6.15 Schemes of p-n junction: **a** respective semiconductors; **b** Formation of p-n junction; **c** charge-carriers transfer when p-n junction is illuminated. Reprinted with permission from Ref. [15]

hole and this makes semiconductor is electrically neutral. On the other hand, the n-type semiconductor contains more electrons than holes, and redundant electron neutralize by fixed positive charge same like the p-type. When p- and n-type semiconductors are combined, due to the carrier concentration in the respective sides the electrons of n region and the hole of p region diffuse to the opposite region under the concentration. Thus the diffused electron into p -region and holes in the n region forms a charge separation layer of certain potential and this region is known as space charge region. Due to the accumulation of charge in both sides, the inner electric field is developed and the direction of the electric field is pointing from n to p region. This inner electric field stops the further movement of carriers driven by concentration difference and a dynamic balance is reached and p–n junction is formed.

When p-type semiconductor and n-type semiconductor contact together to form the p–n junction, a layer of lack of carrier can form in contact point and there are potential barrier to prevent further spread of carrier area, the resistance being very high, known as the barrier layer is also called the depletion layer. As shown in Fig. 6.16, because the area is high resistance area, when on the p–n junction with forward voltage, V makes the potential of p area rise, the hole of p area constantly spread to n area, and the electronic diffuse from n to p, with the thickness of the depletion layer and its built-in electric field intensity decreases, resistivity drop, and current increase. When on the p–n junction with reverse voltage, holes, and electrons were drawn toward each end and flow, and thus the thickness of the depletion layer and built-in electric field strength increases, the resistance will increase, and its through current is greatly reduced. This is the rectifying effect of pn junction [16].

If the photon can get to the p–n junction area, electron-hole pair can be produced on either side of the junction by the intrinsic absorption when the incident light which the photon energy higher than the band gap of the semiconductor irradiate on

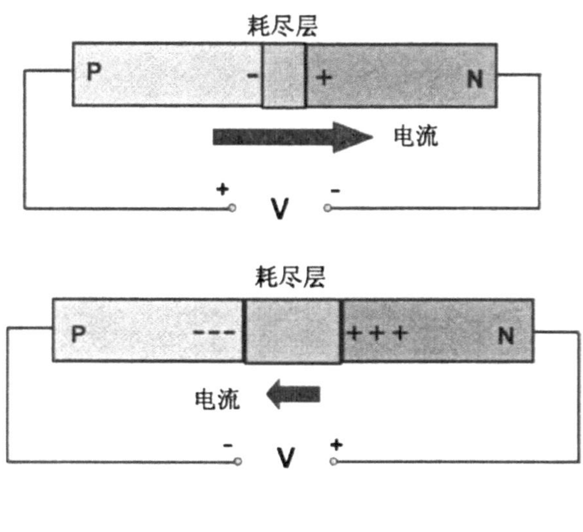

Fig. 6.16 Rectifying effect of p-n junction

p–n junction plane. Under the light excitation, the majority carrier concentration change is very small, minority carrier concentration changed a lot. Due to the built-in electric field (from n to p area) of p–n junction, the minority carrier on both sides of the junction to the opposite direction: electrons of p region through the p–n junction into n, n region hole through the p–n junction into the p region, thus to realize the separation of photogenerated charge. Even a single semiconductor region can be excited under irradiation, but in this case, due to the absence of electric field, photogenerated charges are be removed randomly, leading to electron-hole recombination. Only the photogenerated charge-carriers within the space charge region are driven into the electric field and those space charges outside the space charge region of semiconductor first diffuse into the space charge region at random and then move into the built-in electric field. This mean that those photogenerated charge outside of the space charge region of semiconductor have the same way of movement in the semiconductor as before the spread of the space charge region [17]. Photogenerated electrons driven by the built-in electric field migrate into n region outside the space charge region slowly diffuse in random and as a result the photogenerated electrons continuously come from space charge area due to the difference in the concentration If there was not enough electron acceptor and composite, they would begin to gather resulting in the potential decrease of n region move away from one end of the space charge area; By the same token, the migration and accumulation of the hole toward p region move away from one end of the space charge layer, resulting in higher potential and as a result there is a formation photogenerated potential difference on both ends of the p–n junction. The migration of photogenerated charge was inhibited by photogenerated potential difference, and finally reaches dynamic balance.

Kong et al. [18] synthesized AgBr–BiOBr heterojunction photocatalysts. As indicated in Fig. 6.17, by absorbing visible light the electrons in the VB of the AgBr are excited to its CB leaving the photogenerated holes on VB, and then the electrons in the CB of AgBr can flow down to that of BiOBr, while the

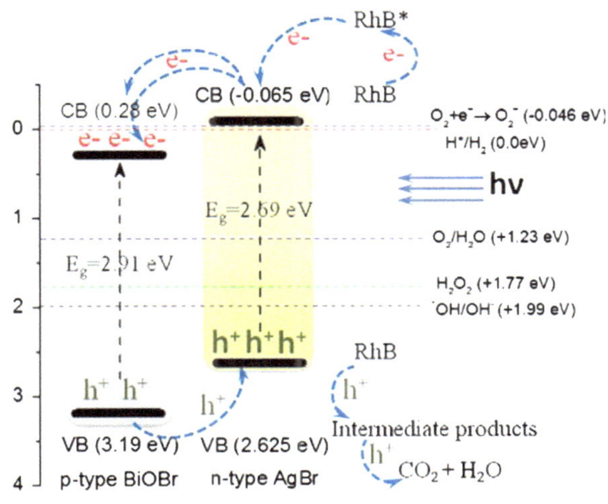

Fig. 6.17 Schematic electronic band diagram of the AgBr–BiOBr system. Reprinted with permission from Ref. [18]

photogenerated positive carriers on VB of BiOBr would transfer to the VB of AgBr. This leads to charge separation and inhibits the recombination of electron-hole pairs. So, AgBr–BiOBr heterojunction photocatalysts exhibited an excellent increase in the photocatalytic activity.

Cheng et al. [19] synthesized AgI/BiOI composite and used it as a photocatalyst and they observed an enhanced photocatalytic activity in the visible region. As shown in Fig. 6.18, a heterojunction is formed between the two semiconducting material and upon visible light excitation, the electrons are excited to the conduction bands of both the semiconductors. Subsequently, the electrons in the conduction band of AgI flows to BiOI and at the same the holes from the valence band of BiOI transfer to that of AgI and establish a spatial separation of charge-carriers. This spatial separation prevents the electron-hole recombination and provides all the charge resources for the photocatalytic reaction.

Cao et al. [20] synthesized the composite of Bi_2S_3 and BiOI and used them as a photocatalyst. As shown in Fig. 6.19, first, the dye as a sensitizer is adsorbed on the photocatalyst, and then is excited to generate electrons under visible light

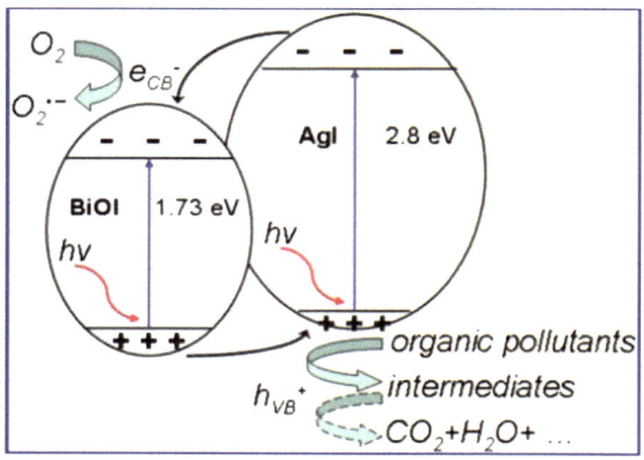

Fig. 6.18 Scheme diagram of the band levels of AgI/BiOI composites and the possible reaction mechanism of the photocatalytic procedure. Reprinted with permission from Ref. [19]

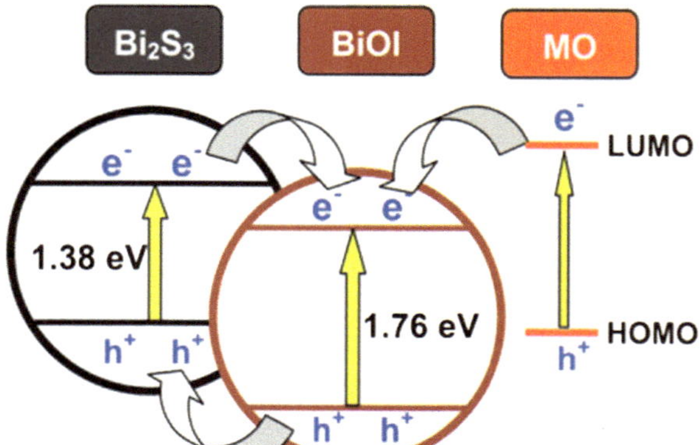

Fig. 6.19 Schematic illustration of the charge transfer pathway during MO degradation process over Bi_2S_3/BiOI under visible light irradiation. Reprinted with permission from Ref. [20]

irradiation; subsequently, these electrons transfer to the CB of the BiOI, and then react with molecular oxygen to form $\bullet O^{2-}$, leaving holes on the VB of dye. On the other hand, the electrons in the VB of the Bi_2S_3 are excited to its CB under absorbing visible light, leaving the photogenerated holes on VB, and then the electrons in the CB of Bi_2S_3 can flow down to that of BiOI; these photogenerated electrons could be transferred to surface-absorbed oxygen to generate activated $\bullet O^{2}$ $^{-}$. The photogenerated positive carriers on VB of BiOI are transferred to the VB of Bi_2S_3. The established charge separation through this process inhibits the recombination of electron–hole pairs.

Ye et al. [21] synthesized $Bi_n(Tu)_xCl_{3n}$/BiOCl nanosheets and used them for phtocatalysis. As indicated in Fig. 6.20, with the excitation wavelength of 420 nm, the photoexcitation takes place only in the $Bi_n(Tu)_xCl_{3n}$ in which the electros are promoted to the conduction band and further moves down to the conduction band of BiOCl. This electron upon reacting with the self absorbed oxygen and generates activated $\bullet O^2$. Generated $\bullet O^{2-}$ and photogenerated holes effectively oxidize RHB.

Shenawi-Khalil et al. [22] prepared BiOBr/Bismuth Oxyhydrate photocatalyst. This is another typical example where the charge transfers take place between the conduction bands and the valence bands of the composite semiconducting materials. Here both BiOBr and Bismuth Oxyhydrate get excited by the radiation and the electrons are promoted to their respective conduction bands. As indicated in Fig. 6.21, the electrons in the CB of BHO flow down to that of BiOBr; these photogenerated electrons react with oxygen to generate activated $\bullet O^{2-}$, while the photogenerated holes on VB of BiOBr transfer to the VB of BHO. This mechanism promotes the charge separation and inhibits the recombination of electron-hole pairs.

Li et al. [23] synthesized BiOCl/BiOI composites. In this case Rhodomine B dye is used as a sensitizer, where the dye on the photocatalyst absorbs visible radiation

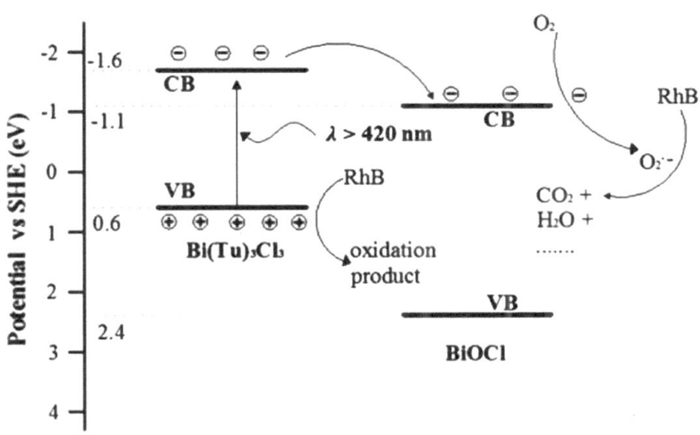

Fig. 6.20 The photocatalytic mechanism of BiOCl nanosheets under visible light irradiation. Reprinted with permission from Ref. [21]

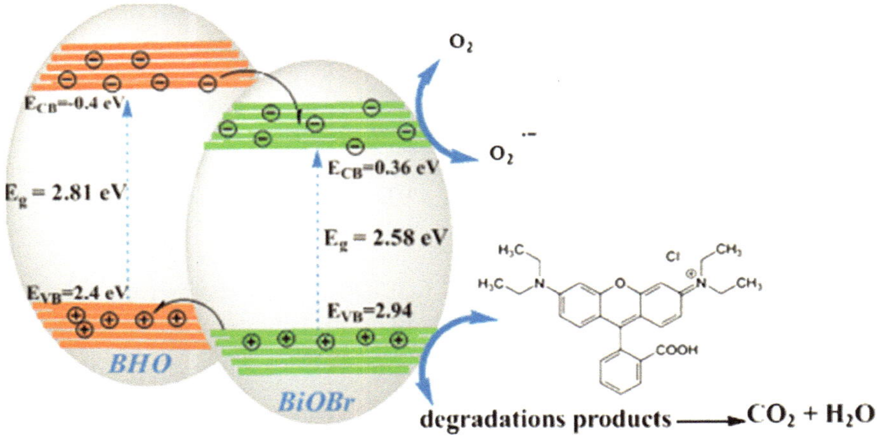

Fig. 6.21 The photocatalytic mechanism of BiOBr/Bismuth Oxyhydrate Photocatalyst under visible light irradiation. Reprinted with permission from Ref. [22]

Fig. 6.22 The proposed possible mechanism for the improvement of photocatalytic activity. Reprinted with permission from Ref. [23]

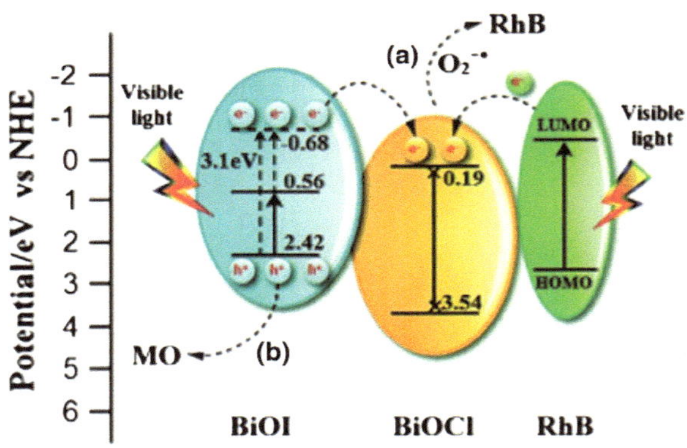

and generated electron and these electrons are transferred to the conduction band of BiOCl and then react with molecular oxygen to form $\bullet O^{2-}$ as shown in Fig. 6.22. On the other hand, the electrons in the VB of the BiOI are excited to its CB under absorbing visible light, leaving the photogenerated holes on VB, and then the electrons in the CB of BiOI can flow down to that of BiOCl; these photogenerated electrons also helps in generating activated $\bullet O^{2-}$. This indicates that the charge separation established in the heterojunction is instrumental in inhibiting the charge recombination and promoting the photocatalytic activity.

Dai et al. [24] prepared BiOI/TiO_2 composites. TiO_2 is n-type semiconductor and BiOI is p-type semiconductor. Before contact, both the conduction band edge and Fermi level of BiOI is lower than that of TiO_2, as shown in Fig. 6.23a. After contact, the Fermi level of TiO_2 is lowered down and the Fermi level of BiOI is

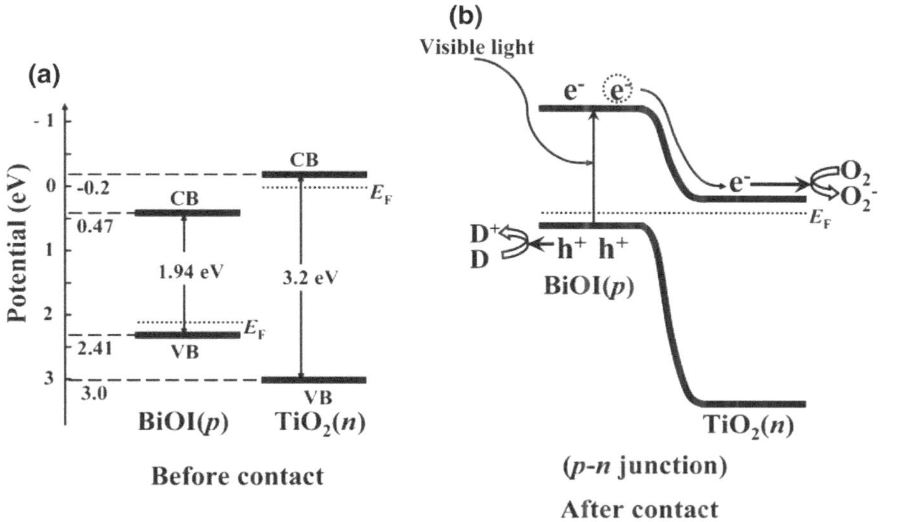

Fig. 6.23 Schematic diagrams for **a** energy bands of p-BiOI and n-TiO_2 before contact, **b** the formation of a p-n junction after contact and the photocatalytic mechanism of BiOI/TiO_2 composites under visible light irradiation. Reprinted with permission from Ref. [24]

raised up until form equilibrium state, as indicated in Fig. 6.23b. As a result, the conduction band edge of TiO_2 is lower than that of BiOI. The electrons in the VB of the BiOI are excited to its CB under absorbing visible light, leaving the photogenerated holes on VB, and then the electrons in the CB of BiOI can flow down to that of TiO_2, these photogenerated electrons could be transferred to surface-absorbed oxygen to generate activated $\bullet O^{2-}$. That promotes charge separation and inhibits the recombination of electron-hole pairs. Generated $\bullet O^{2-}$ and photogenerated holes form could effectively oxidizes the pollutants or organics.

Cao et al. [25] synthesized BiOI/$(BiO)_2CO_3$ catalyst. $(BiO)_2CO_3$ is n-type semiconductor and BiOI is p-type semiconductor. As in the earlier case, the conduction band edge and Fermi level of BiOI is lower than that of $(BiO)_2CO_3$ before they come in contact. After contact, the Fermi level of $(BiO)_2CO_3$ moves down and the Fermi level of BiOI goes up to form a equilibrium state as shown in Fig. 6.24 and this lead up to the conduction band edge of $(BiO)_2CO_3$ to become lower that of BiOI. The electrons in the VB of the BiOI are excited to its CB under absorbing visible light, leaving the photogenerated holes on VB, and then the electrons in the CB of BiOI can flow down to that of $(BiO)_2CO_3$; these photogenerated electrons could be transferred to surface-absorbed oxygen to generate activated $\bullet O^{2-}$. On the other hand, the photogenerated holes on VB of $(BiO)_2CO_3$ goes to the VB of BiOI, that promotes charge separation and inhibits the recombination of electron-hole pairs. Generated $\bullet O^{2-}$ and photogenerated holes form could effectively oxidized the organics.

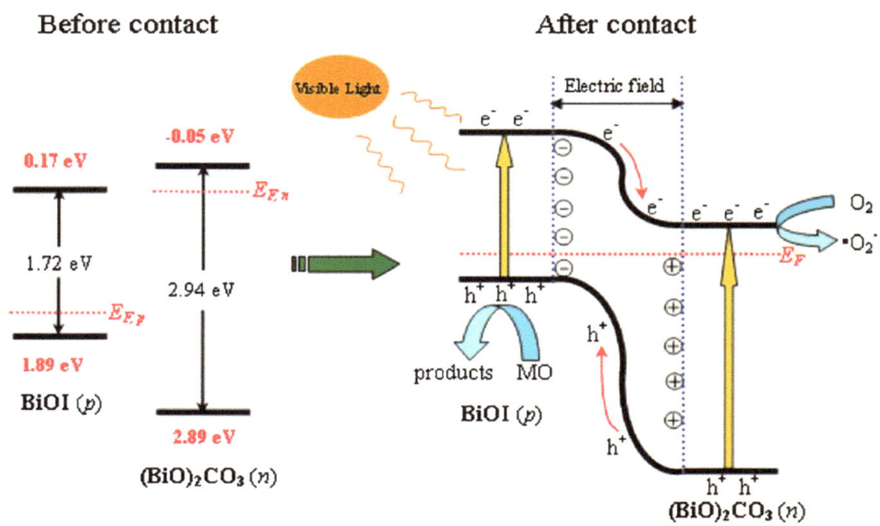

Fig. 6.24 Schematic diagram of the photocatalytic mechanism of BiOI/(BiO)$_2$CO$_3$ composites under visible light irradiation. Reprinted with permission from Ref. [25]

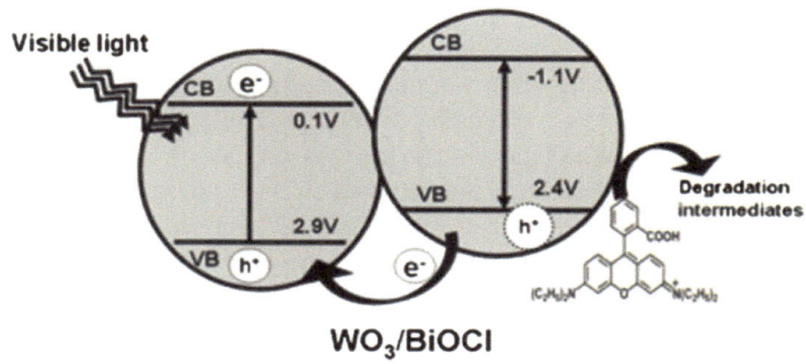

Fig. 6.25 Schematic diagram of the photocatalytic mechanism for the degradation of RHB over WO$_3$/BiOCl under visible light. Reprinted with permission from Ref. [26]

Shamaila et al. [26] synthesized composite of WO$_3$/BiOCl and used this semiconductor composite as a phtocatalyst. The electrons in the VB of the WO$_3$ are excited to its CB under absorbing visible light as shown in Fig. 6.25, and these photogenerated electrons reacts with oxygen to generate activated $\bullet O^{2-}$, and finally the activated $\bullet O^{2-}$ give rise to •OH. Meanwhile, holes are generated in the VB of BiOCl as the electrons from the VB of BiOCl are also transferred to VB of WO$_3$, and this reacts with H$_2$O to generate •OH. Generated •OH oxidizes the organics.

Jiang et al. [27] synthesized ZnO/BiOI heterostructures where ZnO is a n-type semiconductor and BiOI is a p-type semiconductor. Before they are brought into

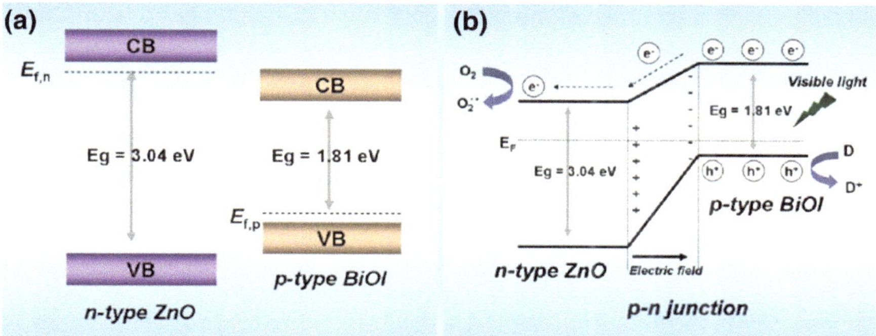

Fig. 6.26 Schematic diagrams for **a** energy bands of n-ZnO and p-BiOI before contact, **b** the formation of a p-n junction after contact and the photocatalytic mechanism of ZnO/BiOI composites under visible light irradiation. Reprinted with permission from Ref. [27]

Fig. 6.27 Schematic diagram for energy band matching and flow of electrons for $BiOCl/Bi_2O_3$ systems under visible light irradiation. Reprinted with permission from Ref. [28]

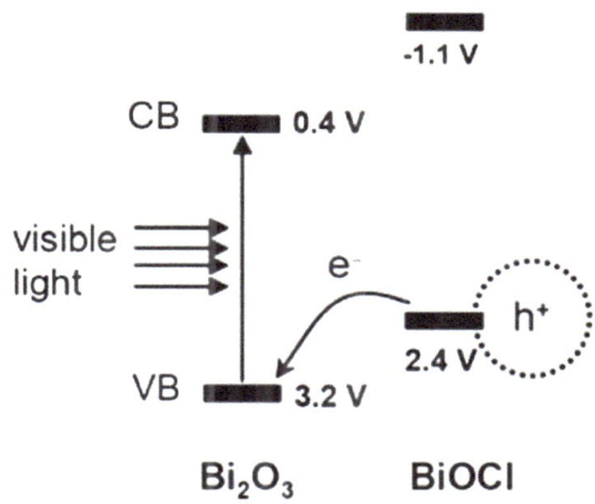

contact, the conduction band edge and Fermi level of BiOI are lower than that of ZnO, as shown in Fig. 6.26a. Once they are in contact, the Fermi level of ZnO is moved down and that of BiOI moves up in order to form a equilibrium state as indicated in Fig. 6.26b. As a result, the conduction band edge of ZnO is lower than that of BiOI and the electrons in the VB of the BiOI are excited to its CB with the absorption of visible light, and then the electrons in the CB of BiOI can flow down to that of ZnO. These photogenerated electrons are transferred to surface-absorbed oxygen to generate activated $\cdot O^{2-}$. The charge separation thus achieved inhibits the recombination of electron and holes and makes all the photogenerated electrons and holes for the redox reaction and thereby enhances the photocatalytic activity.

Chai et al. [28] synthesized $BiOCl/Bi_2O_3$ heterojunction structure where the VB level of BiOCl is higher by 0.7 V than that of Bi_2O_3. Hence, this system is a kind of B-type heterojunction. In this hybrid system, BiOCl, with a band gap of 3.6 eV works as the main photocatalyst, while the role of Bi_2O_3, with a band gap 2.6–2.8 eV, is the sensitizer to absorb visible light. As indicated in Fig. 6.27 by absorbing visible light, the electrons in the VB of the Bi_2O_3 are excited to its CB leaving the photogenerated holes on VB of the Bi_2O_3, and then the electrons in the VB of BiOCl can flow down to that of Bi_2O_3 as its VB is positioned higher than that of Bi_2O_3; finally, photogenerated holes are generated in the VB of BiOCl. Therefore, photocatalytic oxidation reactions can occur on VB of BiOCl; as a result, the B-type heterojunction structure can induce complete mineralization of organic pollutant.

References

1. Linsebigler AL, Lu G, Yates JT (2012) Photocatalysis on TiO_2 surfaces: principles, mechanisms, and selected results. Chem Rev 95:735–758
2. Liu SX, Qu ZP, Han XW (2004) A mechanism for enhanced photocatalytic of siver-loaded titanium dioxide. Catal Today 93–95:877–884
3. Willets KA, Duyne RV (2007) Localized surface plasmon resonance spectroscopy and sensing. Ann Rev Phys Chem 58:267–297
4. Oldenburg SJ, Averitt RD, Westcott SL (1998) Nanoengineering of optical resonances. Chem Phys Lett 28:243–247
5. Schwartzberg AM, Zhang J (2008) Novel optical properties and emerging applications of metal nanostructures. J Phys Chem C 112:10323–10337
6. Sung-Suh HM, Choi JR, Hah HJ (2004) Comparison of Ag deposition effects on the photocatalytic activity of nanoparticulate TiO_2 under visible and UV light irradiation. J Photoch Photobio A Chem 163:37–44
7. Liu H, Cao WR, Su Y (2012) Synthesis, characterization and photocatalytic performance of novel visible-light-induced Ag/BiOI. Appl Catal B Environ 111–112:271–279
8. Yu CL, Yu JC, Fan CF (2010) Synthesis and characterization of Pt/BiOI nanoplate catalyst with enhanced activity under visible light irradiation. Mater Sci Eng B 166:213–219
9. Ai ZH, Ho WK, Lee SC (2011) Efficient visible light photocatalytic removal of NO with BiOBr-graphene nanocomposites. J Phys Chem C 115:25330–25337
10. Su MH, He C, Zhu LF (2012) Enhanced adsorption and photocatalytic activity of BiOI–MWCNT composites towards organic pollutants in aqueous solution. J Hazard Mater 229–230:72–82
11. Zhu SB, Xu TG, Fu HB (2007) Synergetic effect of Bi_2WO_6 photocatalyst with C60 and enhanced photoactivity under visible irradiation. Environ Sci Technol 41:6234–6239
12. Gao FD, Zeng DW, Huang QW (2012) Chemically bonded graphene/BiOCl nanocomposites as high-performance photocatalysts. Phys Chem Chem Phys 14:10572–10578
13. Song S, Gao W, Wang X et al (2012) Microwave-assisted synthesis of BiOBr/graphene nanocomposites and their enhanced photocatalytic activity. Dalton Trans 41:10472–10476
14. Cheng HF, Huang BB, Qin XY (2012) A controlled anion exchange strategy to synthesize Bi_2S_3 nanocrystals/BiOCl hybrid architectures with efficient visible light photoactivity. Chem Commun 48:97–99
15. Tzolov MB, Kuo TF, Straus DA (2007) Carbon nanotube–silicon heterojunction arrays and infrared photocurrent responses. J Phys Chem C 111:5800–5804
16. Xu PS, Sun YM, Shi CS (2003) The electronic structure and spectral properties of ZnO and its defects. Nucl Instrum Methods Phys Res Sect B 199:286–290
17. Patolsky F, Timko BP, Yu G (2006) Detection, stimulation, and inhibition of neuronal signals with high-density nanowire transistor arrays. Science 313:1100–1104
18. Kong L, Jiang Z, Henry H (2012) Unusual reactivity of visible-light-responsive AgBr–BiOBr heterojunction photocatalysts. J Catal 293:116–125

19. Cheng HF, Baibiao BB, Dai Y (2010) One-step synthesis of the nanostructured AgI/BiOI composites with highly enhanced visible-light photocatalytic performances. Langmuir 26:6618–6624
20. Cao J, Xu BY, Lin HL (2012) Novel heterostructured Bi_2S_3/BiOI photocatalyst: facile preparation, characterization and visible light photocatalytic performance. Dalton Trans 41:11482–11490
21. Ye LQ, Gong CQ, Liu JY (2012) $Bi_n(Tu)_xCl_{3n}$: a novel sensitizer and its enhancement of BiOCl nanosheets' photocatalytic activity. J Mater Chem 22:8354–8360
22. Shenawi-Khalil S, Uvarov V, Fronton S (2012) A novel heterojunction BiOBr/Bismuth oxyhydrate photocatalyst with highly enhanced visible light photocatalytic properties. J Phys Chem C 116:11004–11012
23. Li TB, Chen G, Zhou C (2011) New photocatalyst BiOCl/BiOI composites with highly enhanced visible light photocatalytic performances. Dalton Trans 40:6751–6758
24. Dai GP, Yu JG, Liu G (2011) Synthesis and enhanced visible-light photoelectrocatalytic activity of p-n junction BiOI/TiO_2 nanotube arrays. J Phys Chem C 115:7339–7346
25. Cao J, Li X, Lin HL (2012) In situ preparation of novel p-n junction photocatalyst BiOI/$(BiO)_2CO_3$ with enhanced visible light photocatalytic activity. J Hazard Mater 239–240:316–324
26. Shamaila S, Khan A, Sajjad L (2011) WO_3/BiOCl, a novel heterojunction as visible light photocatalyst. J Colloid Interf Sci 356(2011):465–472
27. Jiang J, Zhang X, Sun PB (2011) ZnO/BiOI heterostructures: photoinduced charge-transfer property and enhanced visible-light photocatalytic activity. J Phys Chem C 115:20555–20564
28. Seung YC, Yong JK, Myong HJ (2009) Heterojunctioned BiOCl/Bi_2O_3, a new visible light photocatalyst. J Catal 262:144–149

Chapter 7
Applications of Bismuth Oxyhalides

Abstract This chapter presents some important applications of bismuth oxyhalides, particularly the material's application as a photocatalyst in the process of water purification and water splitting. Other applications covered in this chapter are the application of bismuth oxyhalides in gas sensing, in secondary batteries, and in solar cells.

Keywords Photo-catalyst · Water splitting · Water purification · Gas sensors · Solar cell · Secondary batteries

7.1 Application of BiOX as a Photocatalyst

7.1.1 Degradation of Organic, Inorganic Pollutants, and Microorganisms

One of the major applications of BiOX is their function as photocatalyst in the degradation of organic/inorganic pollutants and deactivation of the microorganisms present in the water and the degree of this photocatalytic activity is usually evaluated by degradation rate of pollutants or inactivation efficiency of pathogens. The most common pollutants present in drinking water are noble metallic ions like Cr^{VI}; organic dyes like tetrabromobisphenol A, methyl orange and neutral red (NR); noxious gas like NO and 2-propanol; and pathogens like *Micrococcus lylae* and *Escherichia coli*. The efficiency of the photocatalyst like BIOX in the degradation process can be evaluated and their sole photocatalytic reaction can be singled out with a series of experiments. These experiments include the deactivation reaction test with and without catalyst in the presence of UV and visible radiation and also with and without radiation in the presence of catalysts. These tests are very important to distingue performance of the synthesized BIOX in the pure photocatalytic degradation process from some non photocatalytic processes that might take place. This is because some of the organic pollutants and microbes are nominally removed and deactivated simply by the UV-visible exposure and also by

M.A. Gondal et al., *Novel Bismuth-Oxyhalide-Based Materials and Their Applications*, Advanced Structured Materials 76,
DOI 10.1007/978-81-322-3739-6_7

coming in contact with some chemicals. For each experiment, the change in the pollutant and pathogen concentration from the initial value is observed and based on these the photocatalytic efficiency of the prepared catalyst can be evaluated.

Zhang et al. [1] synthesized BiOBr microspheres through an ionothermal synthesis route and applied this material to deactivate *M. lylae* pathogen. From Fig. 7.1, it is quite clear that in the absence of UV-visible radiation, the degradation of the pathogen is almost zero, indicated by the 100% survival of *M. lylae*. The other two degradation curves in Fig. 7.1 indicate the *M. lylae* deactivation efficiency using BiOBr nanosheets and BiOBr microspheres in the presence of UV-visible radiation. It is quite clear from Fig. 7.1 that nearly 70% of *M. lylae* could survive after 360 min of visible light irradiation in the presence of BiOBr nanosheet, whereas only 50% of *M. lylae* could survive after 180 min of irradiation and more than 90% deactivation of *M. lylae* is observed after 360 min of visible light irradiation with BiOBr microsphere.

Zhu et al. [2] fabricated Ag/BiOI via a solvothermal process, followed by photodeposition method and in this study, *E. coli* was considered as targeted pathogen. The *E. coli* degradation observed in the above study is depicted in Fig. 7.2, in terms of log of concentration in the ordinate axis for different silver concentration in BiOI. We can observe from Fig. 7.2 that with pure BiOI, the *E. coli* concentration could reduce to only 3.1 in the designated scale after 60 min of visible light exposure, whereas with the optimum concentration of silver in BiOX (8% Ag in BiOI), the *E. coli* deactivation can reach as low as 0.5 scale after 10 min of visible irradiation. Also in this figure we can observe that as the silver concentration in the BiOX increases, the efficiency of deactivation of *E. coli* becomes better, which indicates that by deposition of Ag on BiOI shows an excellence effect about disinfection of *E. coli*.

Liang et al. [3] prepared BiOI–AgI nanocomposites by a solvothermal process and subsequent in situ ion exchange reaction and used this nanocomposite to study the deactivation of *E. coli* microorganism. As it is clear from Fig. 7.3 that neither

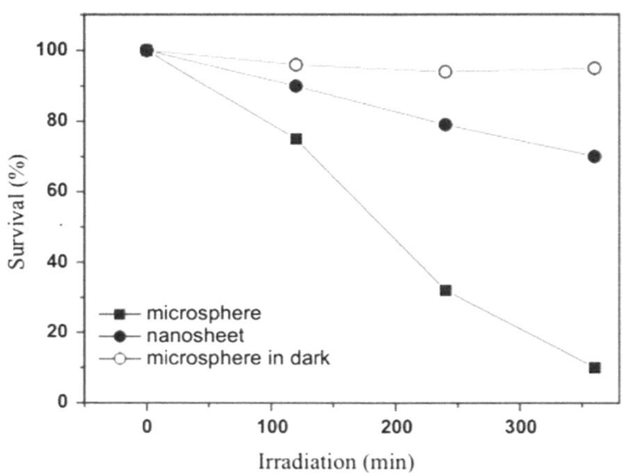

Fig. 7.1 Survival ratio of *M. lylae* versus irradiation time for BiOBr microspheres. Reprinted with permission from Ref. [1]

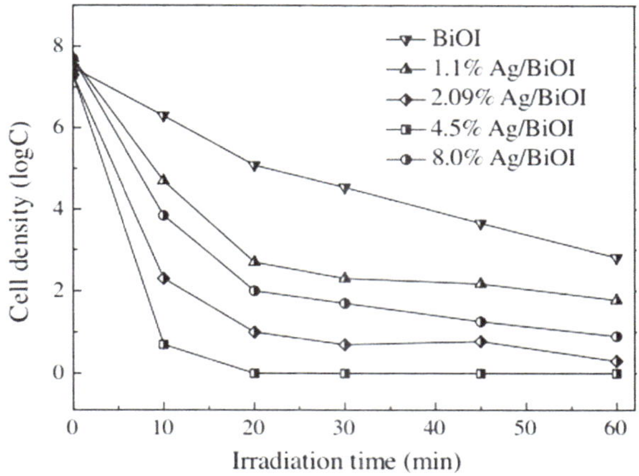

Fig. 7.2 Disinfection efficiencies of *E. coli* versus irradiation time for different Ag content on photocatalytic. Reprinted with permission from Ref. [2]

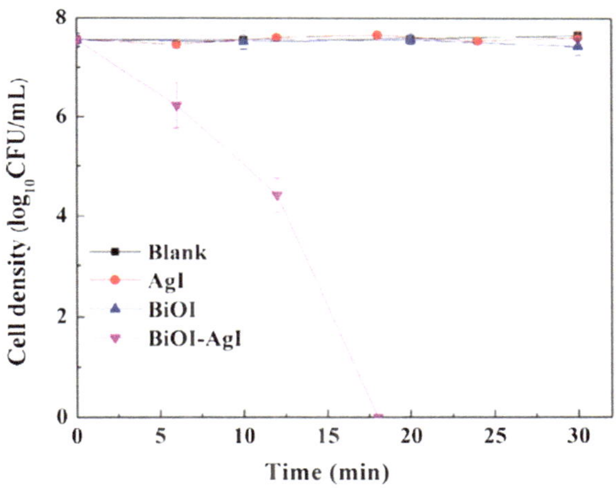

Fig. 7.3 Disinfection efficiencies of *E. coli* by different photocatalysts under visible light irradiation. Reprinted with permission from Ref. [3]

BiOI nor AgI alone in the presence of visible radiation could not bring about any deactivation of *E. coli*, whereas the composite of BiOI–AgI dramatically enhanced the *E. coli* deactivation, which is a typical example of the role of any composite material in any photocatalytic process. Also in Fig. 7.3 we can notice that the BiOI–AgI composite in the absence of visible radiation fail to kill *E. coli*, which clearly indicates the process is purely photocatalytic just due to the formation of heterojunction between BiOI and AgI and consequent reduction in the electron hole recombination.

Wu et al. [4] synthesized two facets (B001 and B010) of BiOBr nanosheets through a facile hydrothermal method and used this material as a photocatalyst to deactivate *E. coli* K-12 microorganism. As it is clear from Fig. 7.4 that both B001 and B010 show no deactivation of *E. coli* K-12 cells in the absence of visible radiation; however, in the presence of light radiation both the materials are proven

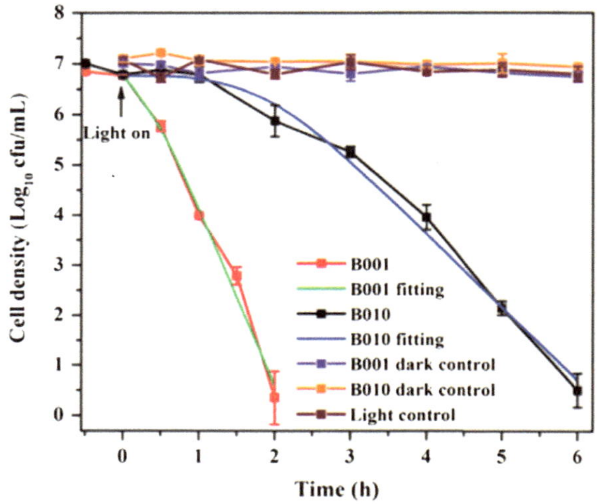

Fig. 7.4 Photocatalytic inactivation efficiency of *E. coli* K-12 in the presence of B001 and B010 nanosheets under VL irradiation. Reprinted with permission from Ref. [4]

to be the good photocatalyst as for as the *E coli* deactivation is concerned. By observing the comparative performance of B001 and B010, for the B001, it takes only 2 h to degrade *E coli* K-12 from the initial value of 7 to 0.5 (described as the log of concentration), where as in the case of B010, in order to achieve the same level of degradation, it takes 6 h. This indicates the phase of the BiOBr plays a vital role in deciding the photocatalytic activity of the material.

In addition to the photocatalytic deactivation of microorganisms, BiOX are used to as photocatalysts for the degradation of organic and inorganic dyes from the water. Gao et al. [5] synthesized BiOCl/Bi_3O_4Cl nanocomposites via a wet-chemical process. In their study, gaseous 2-propanol (IP) was considered as target compound. The degradation curves and the CO_2 gas evolved in the process with BiOCl and Bi_3O_4Cl at various ratios and P25 as a bench mark are shown in Fig. 7.5a, b, respectively. The BiOCl/Bi_3O_4Cl nanocomposite with the optimum 47/53 ratio exhibits the highest catalytic efficiency in terms of the degradation constant *k*. At this optimum ratio of BiOCl/Bi_3O_4Cl nanocomposite, the degradation constant is 2.1 times that of the one achieved with P25 and 6.6 times of the value of k achieved with Bi_3O_4Cl after 2 h of visible light irradiation. By examining Fig. 7.5b, the CO_2 evolved with 47/53 BiOCl/Bi_3O_4Cl was 1.7 times that of P25 and 4 times that of Bi_3O_4Cl with 2 h of visible light irradiation.

Chang et al. [6] applied $NaBiO_3$ for the photocatalytic decomposition of 4-NP, 4-t-OP, and BPA and the compared the results with P25 in Fig. 7.6. As it is clear from Fig. 7.6, $NaBiO_3$ showed an excellent photocatalytic activity achieving 90% of removal compared to mere 9% removal using P25 under visible light or Xenon-lamp irradiation.

Tian et al. [7] fabricated three-dimensional mesoporous BiOBr microspheres through one-step solvothermal process and used it as the photocatalytic removal of bisphenol. As it is clear from Fig. 7.7, BiOBr microsphere is completely incapable

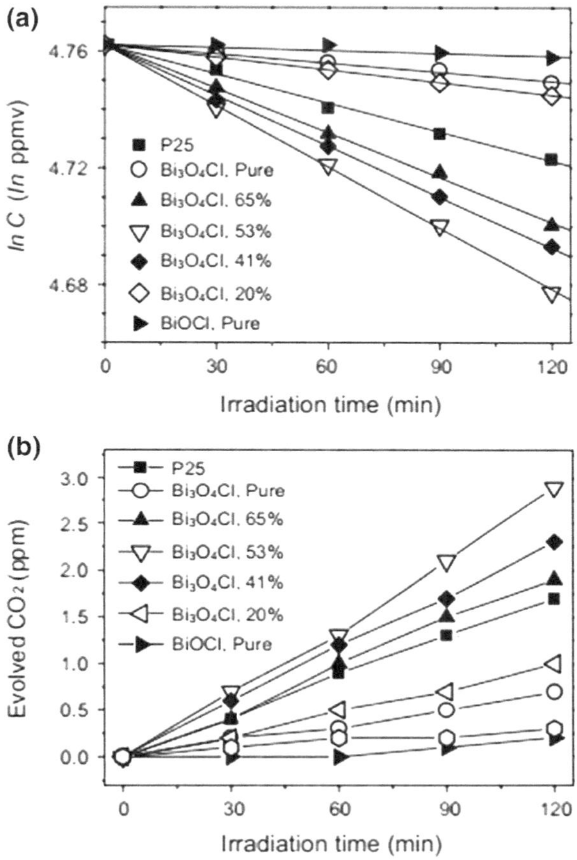

Fig. 7.5 Photocatalytic removal of gaseous IP (**a**) and CO_2 evolved (**b**) with the BiOCl, Bi_3O_4Cl, and the several BiOCl/Bi_3O_4Cl nanocomposites under visible light irradiation. Reprinted with permission from Ref. [5]

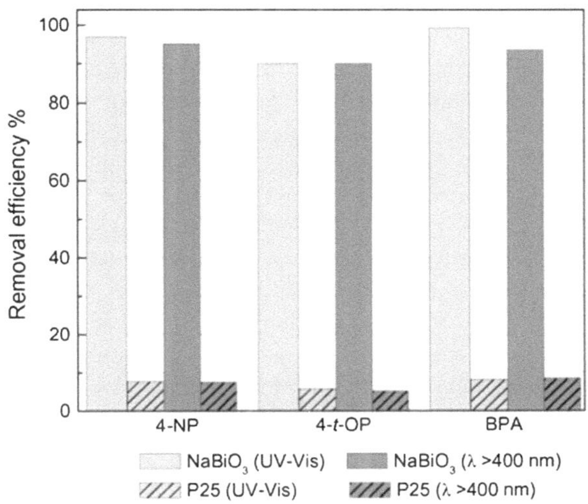

Fig. 7.6 Photocatalytic degradation results of 4-NP, 4-t-OP and BPA comparison between $NaBiO_3$ and P25. Reprinted with permission from Ref. [6]

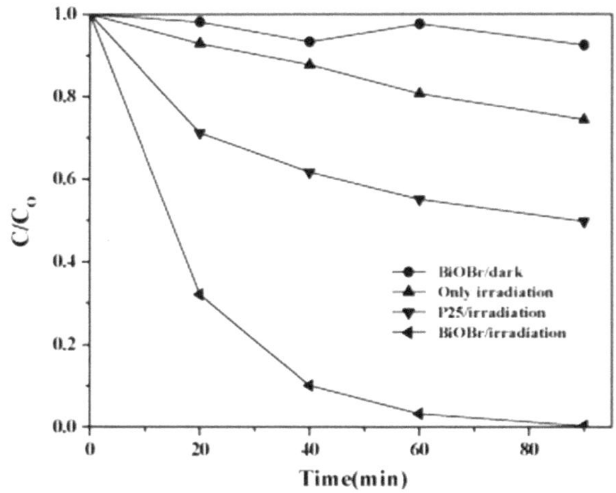

Fig. 7.7 The photocatalytic efficiency of BPA under visible irradiation by BiOBr compared to the results by P25, only with irradiation and BiOBr without irradiation. Reprinted with permission from Ref. [7]

of removing BPA in the absence of light even after 90 min; however, 25.6% degradation of BPA is observed after 90 min of irradiation without any photocatalysts. When BiOBr microspheres is used as a photocatalyst in the presence of light, a 100% BPA degradation was observed after 90 min of irradiation, and on the other hand only 50.3% of BPA could be degraded with P25 as a photocatalyst.

Li et al. [8] synthesized three-dimensional flower-like BiOBr via microwave irradiation method and used for the removal of Cr^{VI} ions clearly indicates that the removal rate of Cr^{VI} ions was about 94% using flower-like BiOBr after 2 h of adsorption in which 79% of Cr^{VI} ions was adsorbed within 15 min and on comparing with plate like BiOBr, a much higher removal efficiency was observed for the flower-like structure.

Qamar et al. [9] synthesized BiOCl and used it as a photocatalyst to remove chromium in the presence of 355 nm laser and a broadband conventional lamp source. As shown in Fig. 7.8, reduction efficiency of chromium is more than 95% within 30 min when BiOCl was used as a catalyst under laser irradiations compared to 49% reduction of chromium achieved using P25 with the same laser irradiation. It is obvious that reduction efficiency of chromium under both laser radiation and conventional UV-lamp using BiOCl is much higher than P25.

Yu et al. [10] synthesized different BiOX by the hydrolysis of BiX_3 (X = Cl, Br) and targeted azo dyes (methyl orange) as model pollutant. Figures 7.9 and 7.10 depict two curves each, one for the ratio of concentration of the pollutant and the other one for the ratio of the total organic carbon (TOC) with BiOCl and BiOBr under UV–VIS illumination, where we can notice that the depletion of concentration of methyl orange is quite faster compared to TOC. This indicates that the incomplete photomineralization of methyl orange molecules.

Ai et al. [11] synthesized BiOBr microspheres via a nonaqueous sol–gel method by using bismuth nitrate and cetyltrimethyl ammonium bromide to remove NO

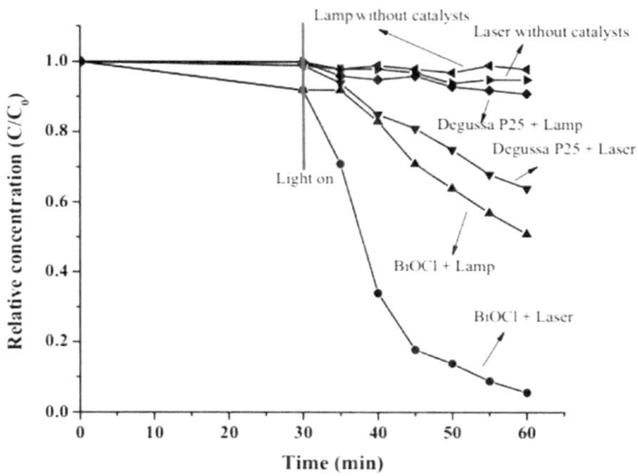

Fig. 7.8 Photocatalytic reduction of chromium under different experimental conditions. Reprinted with permission from Ref. [9]

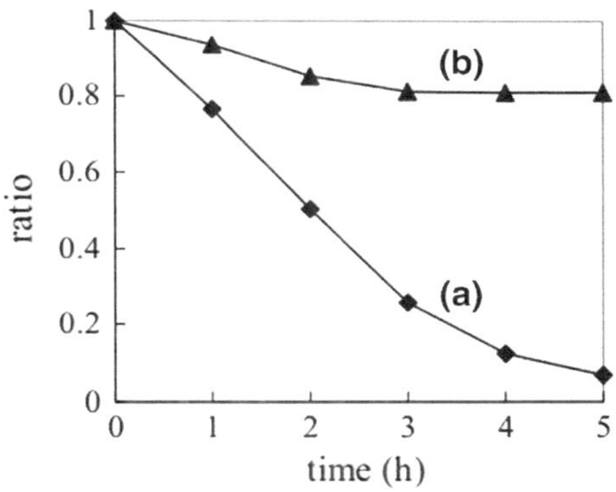

Fig. 7.9 UV–VIS spectral change of MO concentration and TOC value using BiOCl as a function of illumination time, **a** C/Co, **b** TOC/TOCo. Reprinted with permission from Ref. [10]

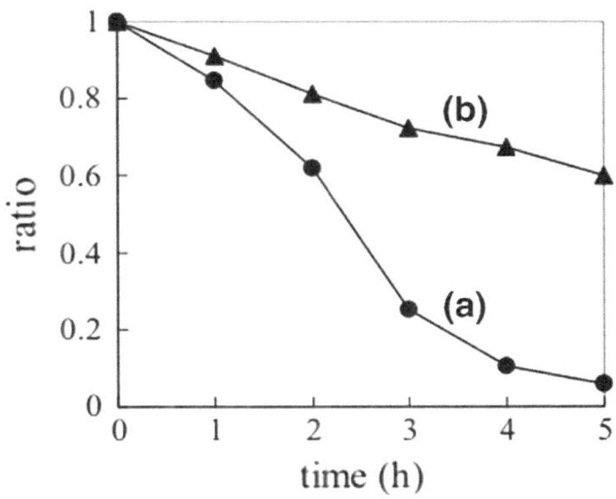

Fig. 7.10 UV–VIS spectral change of MO concentration and TOC value using BiOBr as a function of illumination time, **a** C/Co, **b** TOC/TOCo. Reprinted with permission from Ref. [10]

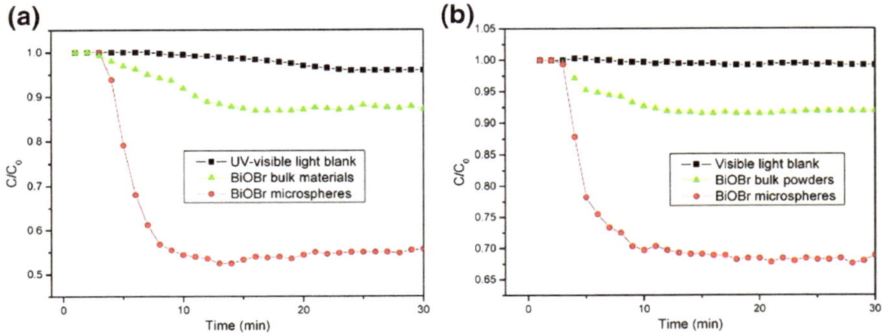

Fig. 7.11 The concentration decrease of NO versus irradiation time in the presence of different photocatalysts. **a** UV-Visible light irradiation, **b** visible light irradiation. Reprinted with permission from Ref. [11]

under visible light irradiation. Figure 7.11a clearly exhibit that the degradation rate of NO was 45% within 10 min when the nonaqueous sol–gel synthesized BiOBr microspheres used as catalyst under UV-visible light, meanwhile, only 13% of NO could be degraded using the chemical precipitation synthesized BiOBr bulk powders in the same conditions. As shown in Fig. 7.11b, it is worth notice that the removal rates was about 30% using the nonaqueous sol–gel synthesized BiOBr microspheres higher than that of 8% for the chemical precipitation synthesized BiOBr bulk powders under visible light irradiation.

Xu et al. [12] synthesized mesoporous BiOBr via solvothermal method and used it for the photocatalytic degradation of tetrabromobisphenol A. Figure 7.12 shows that the removal rate of TBBPA was 48.6% without catalyst after 15 min UV–VIS light irradiation. On the other hand TBBPA was almost completely degraded using BiOBr and on comparing with P25, the degradation rate of BiOBr is almost 75.5% more than that of P25 under the same conditions.

Sarwan et al. [13] used hydrolysis method to prepare BiOCl successfully and NR was used as a model dye for degradation. As shown in Fig. 7.13, the decline in the concentration of NR with BiOCl as a catalyst in the absence of light and the same on the absence of catalyst under mere visible light irradiation is less than 20%. However, the NR degradation rate is approximately 90% with BiOCl, which is almost 40% higher than the same using TiO_2-P25 after 100 irradiation of visible light irradiation.

Feng et al. [14] synthesized mesoporous BiOBr microspheres by solvothermal method and the photodecomposition of gaseous toluene was used to evaluate its photoactivity of the material. As shown in Fig. 7.14, the removal rate of toluene for both P25 and BiOBr were found to be approximately 90% after 300 min of UV irradiation. However, when UV–VIS irradiation was used in the place of UV radiation, the degradation rate of toluene was about 60% with BiOBr used as catalyst compared to 40% degradation using P25 after 300 min of irradiation.

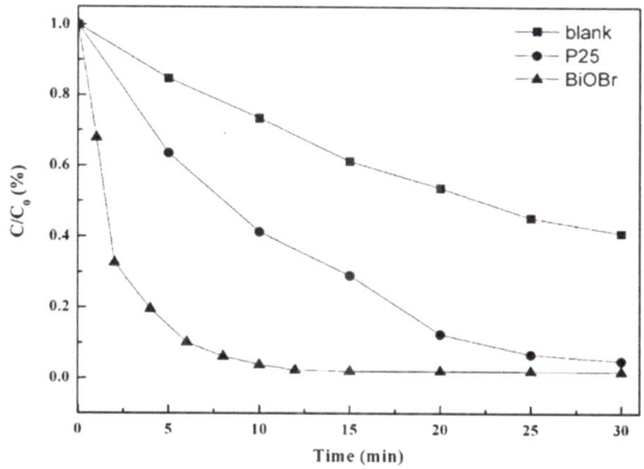

Fig. 7.12 Photocatalytic decomposition of TBBPA by BiOBr and P25 under simulated solar light irradiation. Reprinted with permission from Ref. [12]

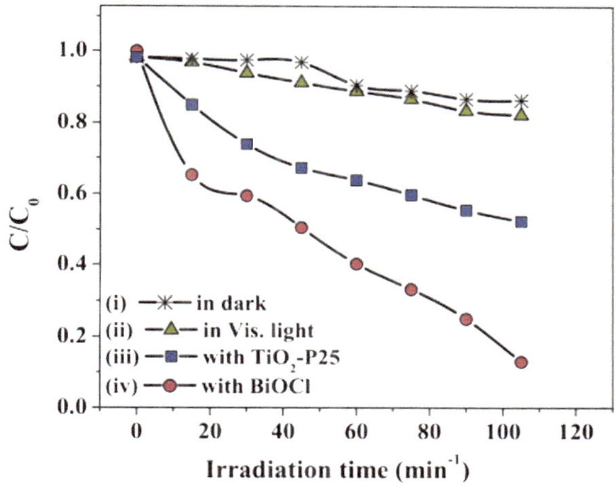

Fig. 7.13 Comparison of change in concentration as a function of irradiation time: (*i*) BiOCl catalyst in dark; (*ii*) only simulated visible light; (*iii*) with TiO_2-P25 catalyst; (*iv*) with BiOCl catalyst under simulated visible light irradiation. Reprinted with permission from Ref. [13]

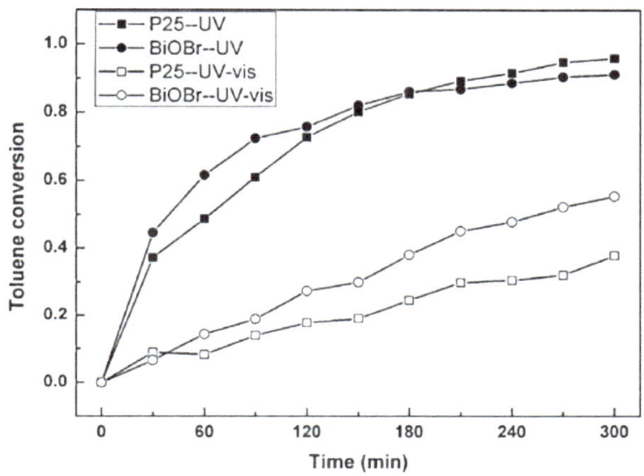

Fig. 7.14 Toluene conversion versus illumination time with P25 and the BiOBr microspheres under various irradiation sources. Reprinted with permission from Ref. [14]

7.1.2 Water Splitting

In a classic photocatalytic water splitting process, water molecules are reduced by photogenerated electrons to form H_2 and oxidized by photogenerated holes to form O_2 leading to overall water splitting. This water spitting is possible with a particular semiconductor photocatalyst when the bottom level of the conduction band is more negative than the redox potential of H $+/H_2$ and the top level of the valence band is more positive than the redox potential of O_2/H_2O. It has been observed that BiOX can be a good photocatalyst meeting the above criterion.

Zhang et al. [15] used BiOCl/CuPc (copper phthalocyanine) composite to study the water splitting reaction and degradation of RhB under different light irradiations. In their study, they observed that when BiOCl/CuPc was used as photocatalyst under the irradiation of LED of 620 nm, RhB was almost completely photodegraded after 2 h, whereas with BiOCl only 34% of RhB could be photodegraded after 2 h. Simultaneously, more O_2 evolution was detected when the RhB and methanol were involved in the reaction system as described in Fig. 7.15.

Zhang et al. [16] investigated that selective transportation of photogenerated electrons and holes in the different facets of BiOCl in the process of pure water splitting. The photogenerated electrons and holes selectively transported to the {001} facets Au nanoparticles and the {110} facets of MnOx under the light irradiation when different photo-deposited BiOCl sample were used as catalysts. As shown in Fig. 7.16, they found that the highest water oxidation activity was observed for the BiOCl@Au/MnOx sample.

Fan et al. [17] successfully synthesized TiO_2–BiOCl double-layer nanostructure arrays for the photoelectrochemical water splitting. As shown in Fig. 7.17, the rutile NWs@anatase/BiOCl NSs show that the highest photocurrent compared to all of other samples of BiOCl deposition cycles of 60.

Fig. 7.15 Photocatalytic water splitting on BiOCl and BiOCl/CuPc under different conditions. Reprinted with permission from Ref. [15]

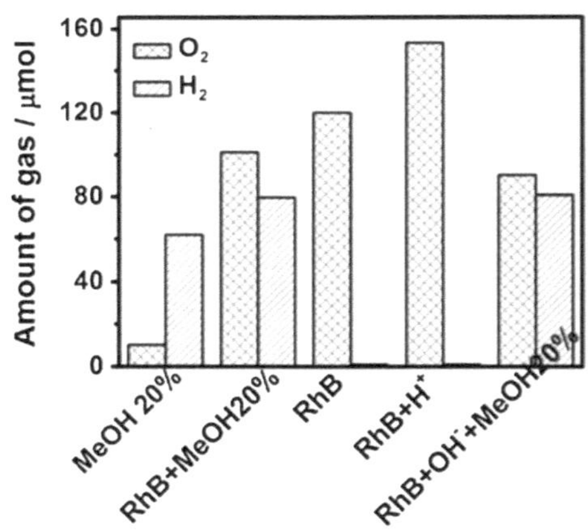

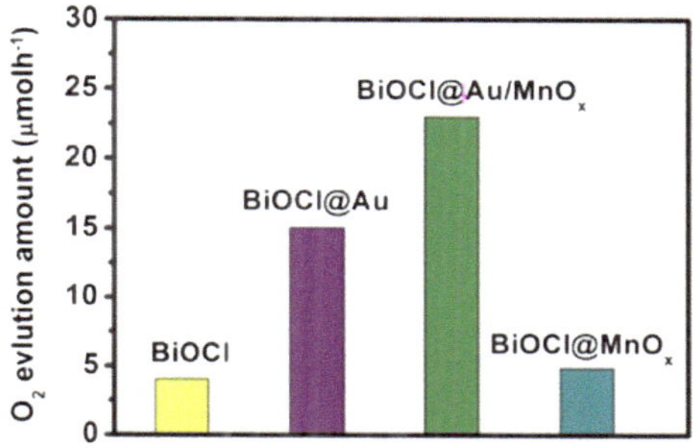

Fig. 7.16 O_2 evolution upon Xe lamp irradiation of an aqueous suspension containing: BiOCl, BiOCl@Au, BiOCl@ MnOx and BiOCl@Au/MnOx, respectively. Reprinted with permission from Ref. [16]

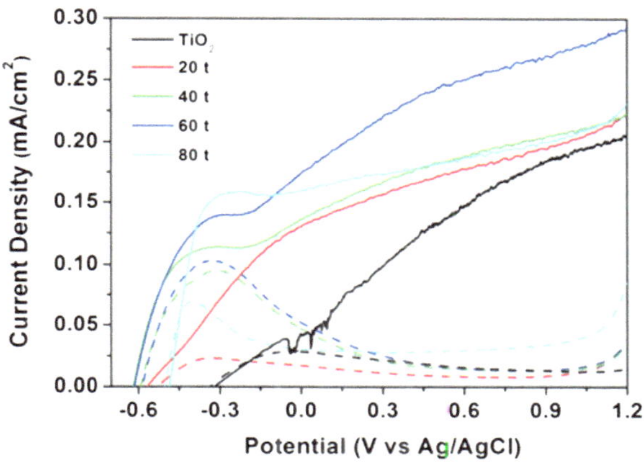

Fig. 7.17 *I–V* characteristic curves of rutile NWs@anatase and rutile NWs@anatase/BiOCl NSs with different number of BiOCl deposition cycles. Reprinted with permission from Ref. [17]

Xian Liu et al. [18] prepared an I-BiOCl/bipolar membrane sandwich structure with photoelectro-synergistic catalysis for water splitting. As shown in Fig. 7.18, when the I-BiOCl/BPM sandwich structure was used as catalyst, their experimental results shows H_2 generation efficiency reached 92 and 35.5% of energy was saved compared to the BPM at a current density of 200 mA cm^{-2} under solar irradiation.

Zhang et al. [19] synthesized BiOCl by reducing the thickness for overall water splitting. In their study, 8.4 μmol of the H_2 was generated from the ultrathin BOC-S after 24 h, and only 2.6 μmol of H_2 was generated by NiOx@BOCS. However, H_2 cannot be detected when the pristine BOC-L and BOC-M were used as photocatalyst for pure water splitting.

7.2 Application of BiOX in Secondary Batteries

Another major application of BiOX is their use in the metal–air batteries. Metal–air batteries are considered to be the most attractive clean energy source due to attractive characteristics like low emissions, portability, high energy density, light

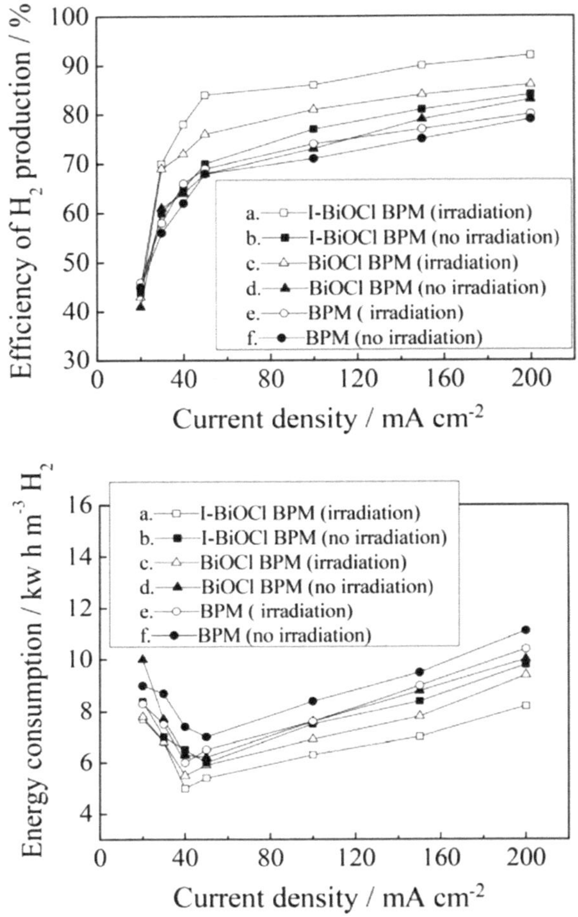

Fig. 7.18 The efficiency and the energy consumption of H_2 generation. Reprinted with permission from Ref. [18]

weight, very low noise, and vibrations. The oxygen reduction reaction is the key for the cathodic reaction taking place in the fuel cell or metal–air battery. The oxygen reduction reactions adopted are a two-electron reduction of O_2 to produce H_2O_2 and the four electron reduction of O_2.

Yuan et al. [20] were successfully synthesized BiOCl micro-assembles by a simple hydrothermal method and they reported that BiOCl (the precursor concentration was 0.18 M) exhibited the working potential of 1.47 eV as shown in Fig. 7.19, which is considered to be the highest working potential window compared to all other samples.

Fujishima et al. [21] developed a Photofuel cell with TiO_2 and BiOCl as photocatalysts and they were able to achieve V_{oc} values of 1.69–1.76 V with the maximum power was 21.1 μW. In their study, the overall efficiency of the cell current per absorbed light was 0.024.

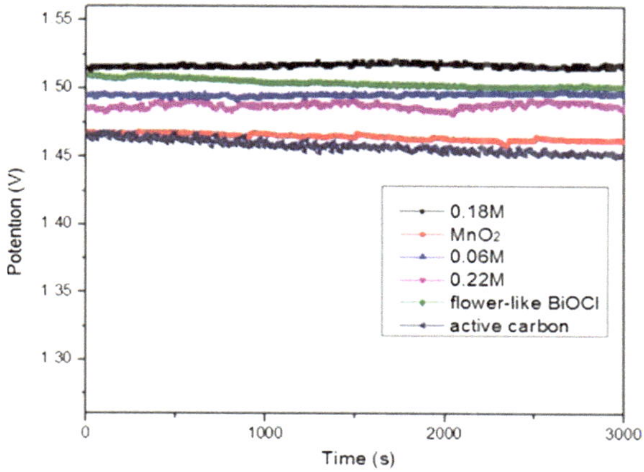

Fig. 7.19 Discharge curve of the air electrode based on the catalyst with the current density of 50 mA cm^{-2}. Reprinted with permission from Ref. [20]

Zhao et al. [22] used metal oxychloride as cathode for the chloride ion batteries. In this study, the FeOCl cathode possesses a discharge capacity of 158 mA h g^{-1} at the first cycle and a stable discharge capacity of 60 mA h g^{-1} in thirtieth cycles.

In addition to metal–air batteries, lithium secondary batteries are the most important power source today. The spinel-type lithium manganese oxides have been studied as important positive electrode materials and the major limitation of this to use them for any practical application is that it has poor cycling performance at high temperatures. In order to overcome these limitations the research has been focused on the possibility of cation replacement at Mn sites, surface modifications, and blending with other materials.

Lee et al. [23] investigated BiOF-coated spinel Li $[Li_{0.1}Al_{0.05}Mn_{1.85}]O_4$ electrode which possesses excellent cycling performance in the voltage range of 3.0–4.3 V at 55 °C and maintain the initial discharge capacity of 96.1% after 100 cycles as shown in Fig. 7.20.

Kanga et al. [24] investigated BiOF-coated 5 V spinel $Li[Ni_{0.5}Mn_{1.5}]O_4$ electrode which possess an excellent cycling performance at 55 °C and maintain capacity retention of 84.5% for 70 cycles compared with initial discharge capacity as in Fig. 7.21.

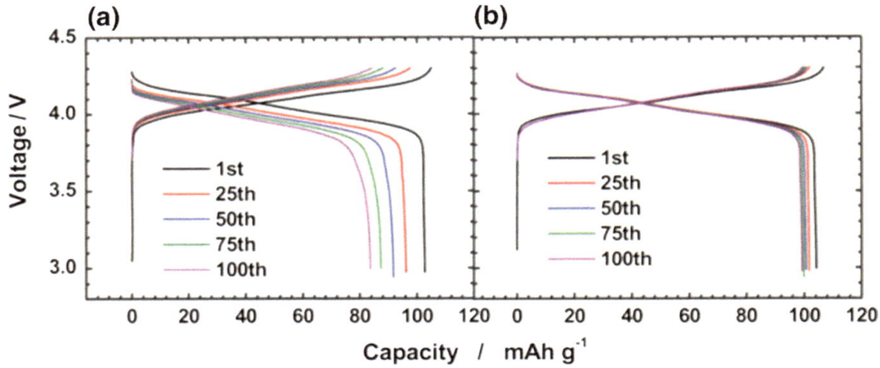

Fig. 7.20 Continuous charge and discharge curves of BiOF-coated spinel $Li[Li_{0.1}Al_{0.05}Mn_{1.85}]O_4$. Reprinted with permission from Ref. [23]

Fig. 7.21 Continuous charge and discharge curves of BiOF-coated 5 V spinel $Li[Ni_{0.5}Mn_{1.5}]O_4$. Reprinted with permission from Ref. [24]

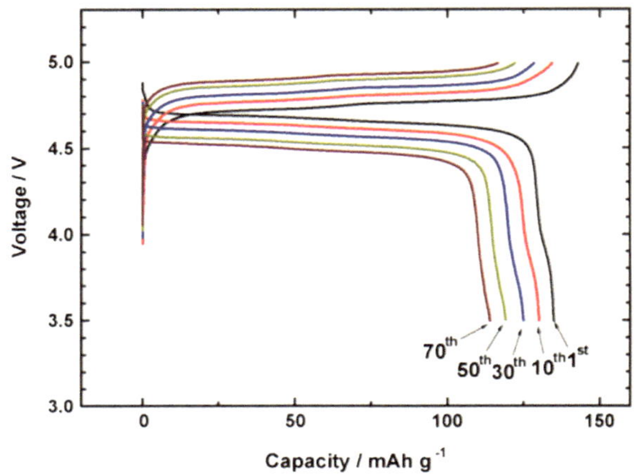

7.3 Application of BiOX in Gas Sensors

Solid state chemical sensors have been developed and put into use to detect and monitor harmful gases such as ozone, SO_2, CO, NO_2, CO_2, and other hazardous gases for several years. Also, some semiconductors are highly sensitive visible light activated photoelectrochemical biosensors for organophosphate pesticide which is an important application in agriculture.

Michel et al. [25] synthesized BiOCl by a surfactant-assisted method and applied them to study the reproducible CO gas sensing patterns at 300 °C. Their work confirmed that a reliable gas sensing patterns can be obtained in all of CO, CO_2, and O_2 gases at 400 °C and it was identified that BiOCl having the characteristics of p-type semiconductor based on the response of BiOCl to oxidizing and reducing gases.

Gong et al. [26] synthesized BiOI nanoflake arrays (BiOINFs) and converted them as a photoactive electrode by a successive ionic layer adsorption and reaction (SILAR). In this study, the detection limit was found to be as low as about 0.04 ng mL^{-1} ($S/N = 3$). Their results show a remarkably improved sensitivity and selectivity for biosensing of OPs with the AChE–BiOINFs/ITO photoactive electrode when experimental conditions are optimum.

7.4 Application of BiOX in Solar Cells

Photovoltaic systems based on semiconductor nanomaterials like dye-sensitized solar cells have been the focus of mainstream research for many years and they herald promising applications in the future. Owing to an excellent photovoltaic response of BiOX, coupled with low cost and low toxicity, these materials have an edge over other semiconductor materials and they could be an excellent potential for photovoltaic applications. Hence, the photovoltaic research centered on BiOX-based solar cells has been in the limelight.

Zhao et al. [27] fabricated BiOI-based solar cells by embedding hierarchical BiOI nanoplate microspheres in chitosan matrix. They find that when the amount of BiOI–CHI colloidal solution was 0.3 mL, the open circuit potential (V_{oc}) was around −461.1 mV, the photocurrent density (I_{sc}) was 20.4 μA/cm^2 at short-circuit conditions, and the fill factor (FF) was 0.46 as depicted in Fig. 7.22.

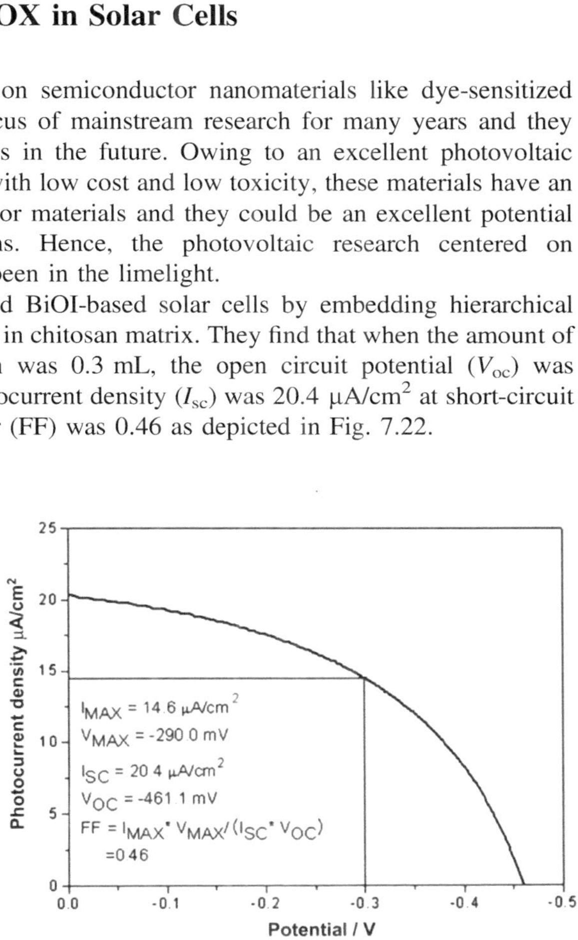

Fig. 7.22 Photocurrent–voltage characteristics of the cell fabricated with 0.3 mL of BiOI-CHI colloidal solution. Reprinted with permission from Ref. [27]

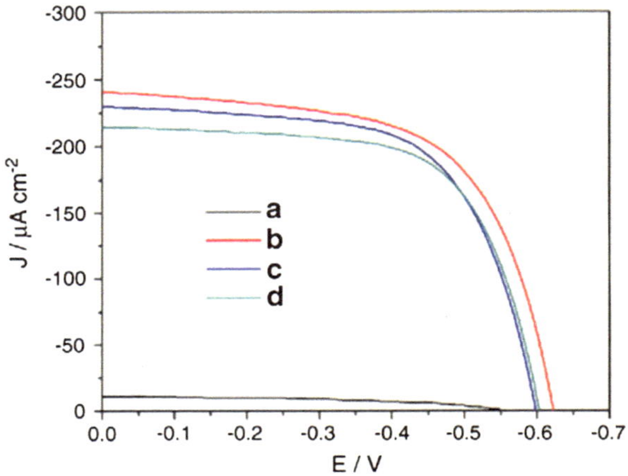

Fig. 7.23 Current–voltage characteristics of BiOI electrodes under 1.5 AM illumination. Annealed TiO_2 block layer/FTO. Reprinted with permission from Ref. [28]

Wang et al. [28] fabricated a solar cell based on crossed flake-like BiOI arrays. A SILAR method was used to fabricate the BiOI flake arrays on an FTO glass with a TiO_2 block layer at room temperature. The resulting figures of merits of such solar cell using BiOI nanoflake array solar cell are as follows where the photocurrent density is 241 μA cm^{-2}, the open circuit potential (V_{oc}) is around 0.62 V, the fill factor is 0.61, the efficiency η is 0.092%, and maximum IPCE is 4% (Fig. 7.23).

Luz et al. [29] prepared BiOCl and BiOBr nanodiscs by water-based nucleation method and purified them through a phase-transfer reaction and the characteristic features and functions of a solar cell with established p-DSSCs were shown in their study. The figure of merit with BiOX-based material are as follows, V_{oc} = 120 mV, J_{sc} = 57 μA cm^{-2}, FF = 40.6%, and η = 0.003 obtained using BiOCl/coumarin 343/KI-I_2; and also the using BiOBr/eosin Y/[C_4MPyr]$_2$[Br_{20}] are V_{oc} = 78 mV, J_{sc} = 3.1 μA cm^{-2}, FF = 28.6%, η = 0.0005 as shown in Fig. 7.24.

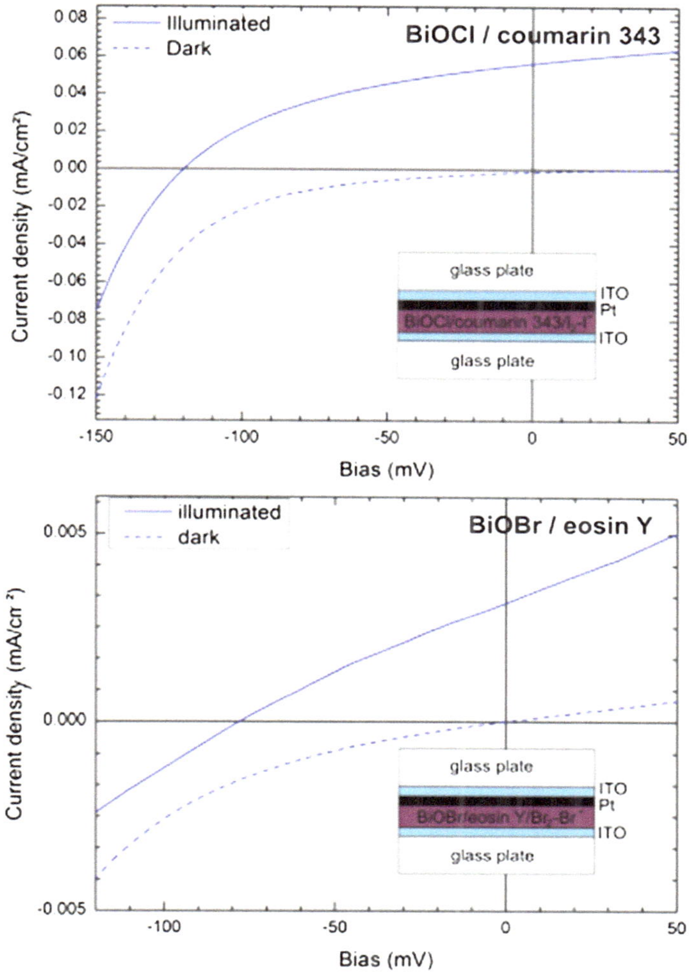

Fig. 7.24 Illustration of BiOCl/coumarin 343 (*top*) and BiOBr/eosin Y (*bottom*) test solar cells and JeV diagrams of the resulting DSSC with KI-I_2 (*top*) and $[C_4MPyr]_2[Br_{20}]$ polybromide (*bottom*) electrolytes under AM1.5 simulated sunlight (*continuous line*) and dark current of the cell. Reprinted with permission from Ref. [29]

References

1. Zhang DQ, Wen MC, Jiang B et al (2012) Ionothermal synthesis of hierarchical BiOBr microspheres for water treatment. J Hazard Mater 211–212:104–111
2. Zhu LF, He C, Huang YL et al (2012) Enhanced photocatalytic disinfection of *E. Coli* 8099 using Ag/BiOI composite under visible light irradiation. Sep Purif Technol 91:59–66
3. Liang JL, Shan C, Zhang X et al (2015) Bactericidal mechanism of BiOI–AgI under visible light irradiation. Chem Eng J 279:277–285

4. Wu D, Wang B, Wang W et al (2015) Visible-light-driven BiOBr nanosheets for highly facet-dependent photocatalytic inactivation of *Escherichia coli*. J Mater Chem A 3: 15148–15155
5. Gao BF, Chakraborty AK, Yang JM et al (2010) Visible-light photocatalytic activity of BiOCl/Bi_3O_4Cl nanocomposites. Bull Korean Chem Soc 31:1941–1944
6. Chang XF, Huang J, Cheng C et al (2010) Photocatalytic decomposition of 4-t-octylphenol over $NaBiO_3$ driven by visible light: catalytic kinetics and corrosion products characterization. J Hazard Mater 173:765–772
7. Tian HT, Li JW, Ge M et al (2012) Removal of bisphenol A by mesoporous BiOBr under simulated solar light irradiation. Catal Sci Technol 2:2351–2355
8. Li GF, Qin F, Yang H et al (2012) Facile Microwave Synthesis of 3D Flowerlike BiOBr Nanostructures and Their Excellent Cr^{VI} Removal Capacity. Eur J Inorg Chem 2012: 2508–2513
9. Qamar M, Yamani ZH (2012) Bismuth oxychloride-mediated and laser-induced efficient reduction of Cr(VI) in aqueous suspensions. Appl Catal A Gen 439–440:187–191
10. Yu ZY, Detlef B, Ralf D et al (2012) Photocatalytic degradation of azo dyes by BiOX (X = Cl, Br). J Mol Catal A Chem 365:1–7
11. Ai ZH, Ho WK, Lee SC et al (2009) Efficient photocatalytic removal of NO in indoor air with hierarchical bismuth oxybromide nanoplate microspheres under visible light. Environ Sci Technol 43:4143–4150
12. Xu J, Meng W, Zhang Y et al (2011) Photocatalytic degradation of tetrabromobisphenol A by mesoporous BiOBr: Efficacy, products and pathway. Appl Catal B Environ 107:355–362
13. Sarwan B, Pare B, Acharya AD et al (2012) Mineralization and toxicity reduction of textile dye neutral red in aqueous phase using BiOCl photocatalysis. J Photochem Photobiol B 116:48–55
14. Feng YC, Li L, Li JW et al (2011) Synthesis of mesoporous BiOBr 3D microspheres and their photodecomposition for toluene. J Hazard Mater 192:538–544
15. Zhang L, Wang WZ, Sun SM et al (2013) Water splitting from dye wastewater: a case study of BiOCl/copper(II) phthalocyanine composite photocatalyst. Appl Catal B Environ 132–133:315–320
16. Zhang L, Wang WZ, Sun SM et al (2015) Selective transport of electron and hole among 001 and 110 facets of BiOCl for pure water splitting. Appl Catal B Environ 162:470–474
17. Fan WQ, Yu XQ, Song SY et al (2014) Fabrication of TiO_2–BiOCl double-layer nanostructure arrays for photoelectrochemical water splitting. CrystEngComm 16:820–825
18. Liu X, Yang HM, Dai HY et al (2015) A novel photoelectrocatalytic approach for water splitting by an I-BiOCl/bipolar membrane sandwich structure. Green Chem 17:199–203
19. Zhang L, Han ZK, Wang WZ et al (2015) Solar-light-driven pure water splitting with ultrathin BiOCl nanosheets. Chem Eur J 21:18089–18094
20. Yuan JL, Wang J, She YY et al (2014) BiOCl micro-assembles consisting of ultrafine nanoplates: a high performance electro-catalyst for air electrode of Al–air batteries. J Power Sources 263:37–45
21. Fujishima Y, Okamoto S, Yoshiba M et al (2015) Photofuel cell comprising titanium oxide and bismuth oxychloride ($BiO_{1-x}Cl_{1-y}$) photocatalysts that uses acidic water as a fuel. J Mater Chem A 3:8389–8404
22. Zhao XY, Zhao-Karger ZR, Wang D et al (2013) Metal oxychlorides as cathode materials for chloride ion batteries. Angew Chem Int Ed 52:13621–13624
23. Lee KS, Myung ST, Amine K et al (2009) Dual functioned BiOF-coated $Li[Li_{0.1}Al_{0.05}Mn_{1.85}]O_4$ for lithium batteries. J Mater Chem 19:1995–2005
24. Kanga HB, Myung ST, Amine K et al (2010) Improved electrochemical properties of BiOF-coated 5V spinel $Li[Ni_{0.5}Mn_{1.5}]O_4$ for rechargeable lithium batteries. J Power Sources 195:2023–2028
25. Michel CR, López Contreras NL, Martínez Preciado AH (2011) Gas sensing properties of nanostructured bismuth oxychloride. Sensors Actuators B 160:271–277

26. Gong JM, Wang XQ, Li X et al (2012) Highly sensitive visible light activated photoelectrochemical biosensing of organophosphate pesticide using biofunctional crossed bismuth oxyiodide flake arrays. Biosens Bioelectron 38:43–49
27. Zhao K, Zhang X, Zhang LZ (2009) The first BiOI-based solar cells. Electrochem Commun 11:612–615
28. Wang KW, Jia FL, Zheng Z et al (2010) Crossed BiOI flake array solar cells. Electrochem Commun 12:1764–1767
29. Andreas L, Jonas C, Michael W et al (2013) p-DSSCs with BiOCl and BiOBr semiconductor and polybromide electrolyte. Solid State Sci 19:172–177

Chapter 8
Latest and Future Prospects of Bismith Oxyhalides

Abstract This chapter mainly describes the future prospect of bismuth oxylalides in terms of their novel preparation techniques, and its possible applications in various other fields. It also describes the possible ways to improve the functionality by synthesizing multidimensional shape dependant materials and heterojunction structures.

Keywords Multidimensional shape dependant materials · Heterojunction structures · Novel synthesis techniques

There are two great issues which world is facing at present due the increase in the environmental pollution and energy shortages as a result of the rapid industrialization development and population growth. Under these circumstances, there is high demand for remediation of environment via development of pollution-free technologies and alternative renewable energy sources. All these issues have attracted extensive research interests by scientists around the world. Over the last few years, a variety of strategies have been investigated to develop efficient and pollution free technologies for the demineralization of highly toxic pollutants to non-toxic products. Under these technologies, photo-catalysis has been emerged as one of the most favorable technology because it embodies more economical and green way to demineralize pollutants by using the energy from sunlight or artificial light using broad spectral lamps for illumination. Since sunlight is abundantly available all over the world so the photo-degradation of highly toxic species using the photocatalytic process holds great potential for the removal of organic(dyes, phenolic compounds, pesticides, herbicides), heavy metals and other toxic pathogens (bacteria, fungus candida) from water [1–74].

M.A. Gondal et al., *Novel Bismuth-Oxyhalide-Based Materials and Their Applications*, Advanced Structured Materials 76,
DOI 10.1007/978-81-322-3739-6_8

8.1 Future Plans for Preparation of BIOX

Ever since the pioneer work reported by Fujishim and Honda in 1972 [75] on the use of TiO_2 for photo-induced water splitting, TiO_2 still has been utilized as a long lasting photo catalyst. However TiO_2 have an inherited problem due to its large band gap (3.2 eV) and so its absorption is only limited to the UV spectral region which is only 4% of the full solar spectrum. Hence scientific community is looking for other alternative semiconductor photo catalysts which could absorb well in the visible region to harvest fully the solar spectrum (sunlight). Under such scenario, development of new highly efficient photo-catalysts has become a hot topic for researchers around the world. Among these catalysts, bismuth oxyhalides have attracted considerable interest due to their efficient photocatalytic activity, high stability, low cost and other properties like magnetic separation after use in water. In recent years, a lot of efforts have been dedicated to synthesize various bismuth oxyhalides using different techniques. These efforts have been focused on those bismuth oxyhalide (BiOX, X= Cl, Br, I) photo-catalysts which are highly efficient, cheap and could absorb in the visible region [76–92]. In this book, we have focused on BiOCl, BiOBr, BiOI. In addition, non metal doping of these BiOX were also reported such as $BiOX_{1-x}\,Y_x$ ($X \neq Y$, X, Y = Cl, Br, I) such as BiOClBr, BiOBrI, BiOClI [88, 91, 93–96]. This method is required to shift the valence band edge to higher energy values and also to enhance the photo-catalytic activity. Recently more focus has been how to use the solar energy to excite such catalyst in the visible region and to achieve such objectives, new scheme for excitation have been introduced in which suitable dyes were used as sensitizer. In such scheme, first, an electron is excited form HOMO state of dye to LUMO state and then such electron is transferred to the conduction band of the BiOX photo-catalysts. It was noticed that BiOX could exhibit high visible-light response photocatalytically when mixed with suitable dyes [81, 82, 86]. This work will continue in future as it is very promising due to its utility of visible light as an excitation source and thus the use solar radiations for future field applications. Some recent work on this issue by our group and others is worth reading.

8.2 Focus on Multidimensional Shape-Dependent (or Layered) BiOX

In addition to above mentioned initiatives, there is more focus recently on the structure of these BiOX photo-catalysts for better crystallinity. The basic unit of nanomaterials can be categorized into three shapes: zero-dimensional (1-D) such as nanoparticles with the three dimensions in nanometer size range; one-dimensional (1-D) such as nanowires, nanofibers, nanorods, nanotubes and nanoribbons with two dimensions at nanoscale; two-dimensional (2-D) such as nanoplates and nanosheets with one-dimensional space at nanoscale. Nanostructured Catalysts based on Bismuth compounds with different dimensionalities including dots, rods, and sheets exhibited clearly dimensionality dependent photocatalytic activity for degradation of various dyes [97]. For example, snow like 1-D Bi_2WO_6 particles showed high photocatalytic removal efficiency for dye under natural sunlight, which was due to its high surface area resulting from the small size and porous structures [98]. Quasi-one dimensional Bi_2WO_6 nanostructures showed efficient photoactivity for the decomposition of RhB under visible light irradiation [99]. Nanoplate-like 2-D Bi_2WO_6 structure with the size of about 30 nm in length performed high photoactivity for degradation of RhB in water under visible light and even a compact fluorescent lamp irradiation, which was attributed to its small particle size and large surface area [20]. Nanostructures with different dimensionalities including dots, rods, and sheets showed dimensionality dependent photocatalytic activity for degradation of rhodamine B, methylene blue and methyl orange [100]. Snow like 2-D Bi_2WO_6 particles exhibited high photocatalytic degradation efficiency for RhB under natural sunlight, which was attributed to its high surface area resulting from the small size and porous structures [99]. The various synthesis processes are described in Fig. 8.1.

Quasi-one dimensional Bi_2WO_6 nanostructures exhibited efficient photoactivity for the decomposition of rhodamine B under visible light irradiation [99]. Nanoplate-like 2D Bi_2WO_6 structure with the size of about 30 nm in length performed high photoactivity for degradation of RhB in water under visible light and even a compact fluorescent lamp irradiation, which was attributed to its small particle size and large surface area [100]. Bi_2WO_6 photocatalysts with nanosheet morphology also displayed high photocatalytic property for RhB degradation under visible light irradiation, which was attributed to the specific surface area and oxygen

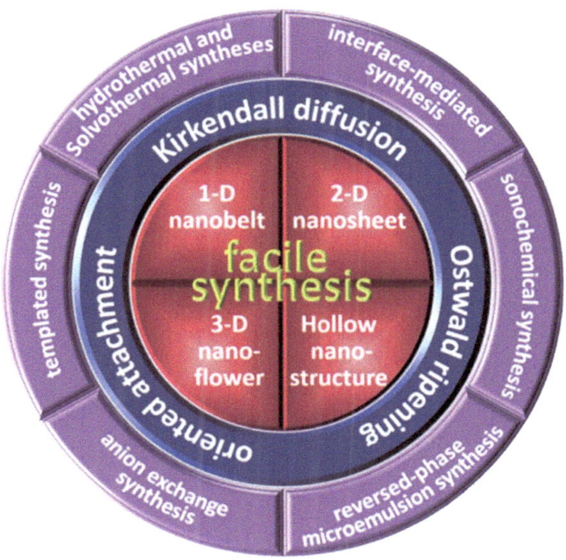

Fig. 8.1 Schematic of the synthesis schemes and growth mechanisms of nanostructured BiOX. Reprinted with permission from Ref. [100]

vacancy [101, 102]. Nanovoid Bi_2WO_6 2D ordered arrays used as a photoanode for water splitting under visible light irradiation was investigated by Bahnemann's group, which performed higher H_2 evolution efficiency than the flat conventional Bi_2WO_6, owing to the reduced reflection losses of the impinging photo energy, large surface area, and the enhanced separation efficiency of the photo-generated charge-carriers resulting from its specific nanovoid 2D array structure [103]. Recently, Chen's group demonstrated that various bismuth-related semiconductors with different dimensionalities and morphologies performed remarkably different photocatalytic activities for dye degradation and heavy metal ions removal. For instance, well crystallized square like 2D BiOCl nanostructures exhibited superior visible light-driven photocatalytic efficiency for RhB degradation than the commercial BiOCl and TiO_2 (anatase), which was mainly attributed to the synergistic effect of the layered structure, the fast charge transfer, and the strong adsorption for RhB dye upon the BiOCl nanoplates [104]. Xiong et al. found that Bi_2O_3 and $(BiO)_2CO_3$ nanotubes displayed much higher Cr(VI)-removal capacity and photocatalytic activity for the degradation of RhB than 2D nanosheets under visible light irradiation [25]. Morphology-dependent photocatalytic performance of $BiPO_4$ nanostructures was also investigated. It was found that hexagonal $BiPO_4$ nanoparticles showed highest photocatalytic performance [1] Bi_2WO_6 photocatalysts with nanosheet morphology also displayed high photocatalytic property for RhB degradation under visible light irradiation, which was attributed to the specific surface area and oxygen vacancy [105]. Nanovoid Bi_2WO_6 2D ordered arrays used as a photoanode for water splitting under visible light irradiation was investigated by Bahnemann's group, which performed higher H_2 evolution efficiency than the flat conventional Bi_2WO_6, owing to the reduced reflection losses of the impinging photo energy, large surface area, and the enhanced separation efficiency of the photo-generated charge-carriers resulting from its specific nanovoid 2D array structure [99]. Recently, Chen's group demonstrated that various bismuth-related semiconductors with different dimensionalities and morphologies performed remarkably different photocatalytic activities for dye degradation and heavy metal ions removal. For instance, well crystallized square like 2D BiOCl nanostructures exhibited superior visiblelight-driven photocatalytic efficiency for RhB degradation than the commercial BiOCl and TiO_2 (anatase), which was mainly attributed to the synergistic effect of the layered structure, the fast charge transfer, and the strong adsorption for RhB dye upon the BiOCl nanoplates. Xiong et al. found that Bi_2O_3 and $(BiO)_2CO_3$ nanotubes displayed much higher Cr(VI)-removal capacity and photocatalytic activity for the degradation of RhB than 2D nanosheets under visible light irradiation [25]. Morphology-dependent photocatalytic performance of $BiPO_4$ nanostructures was also investigated. It was found that hexagonal $BiPO_4$ nanoparticles showed highest photocatalytic performance. For the MO degradation under both UV and visible light irradiation among various $BiPO_4$ nanostructures, which was attributed to its crystalline phase and narrow band gap [105].

8.3 Future Methodologies for the Production of Nano BiOX

Over the past several years, controllable synthesis of the bismuth-related primary nanoscale building blocks such as nanoparticles, nanowires, nanobelts, and nano-plates, have made great progress for photocatalysis. Having square like two dimensional (2D) nanoplates/or sheets or hierarchical three dimensional (3D) nanostructures and such catalysts have been widely developed and applied for environmental remediation [97–99]. These 3D hierarchical structures exhibited visible-light induced photo-catalytical removal of Dyes from water which is well desired by research community and the public at large. The future trends in the development of BiOX seems mostly on development of catalysts like 2D and 3D nanosheets, nanoplates and their applications not only limited for dyes removal from water but heavy metals, pathogens and phenolic compounds from water and also for water splitting to generate hydrogen for clean energy environmental initiatives [97–99].

Recent studies showed that the photo activity of BiOX nanosheets (NS) was greatly influenced by the exposure of the side of surface. The BiOCl with exposed {001} facets under the UV light, showed larger photoexcitation activity for degradation of pollutants because of a supportive effect amongst the internal electric field and the surface atomic structure. BiOCl NS with exposed {010} facet under the visible light obsessed higher dye photosensitization activity for methyl orange degradation due to the bigger surface area and open channel characteristic of BiOCl NS. In this case, the adsorption capacity of methyl orange molecules is increased due to exposed {010} facets which offer more contact sites between the dye molecules and photo-catalyst, which enables electron injection process efficient of the photo excited dye into the catalyst conduction band, which is beneficial for the subsidiary dye photosensitization process as shown in Fig. 8.2. These results elucidated the origin of photo reactivity dependence on facet-properties of BiOCl NS, and provides effective direction for the synthesis of very efficient bismuth oxyhalide photo-catalysts [100].

In the BiOX, photocatalysts family, BiOI has demonstrated the highest absorption efficiency in the visible region thus seems to be most potential candidate for solar light harvesting but it has intrinsic problem due to rapid electron hole recombination. Due to this reason, single component BiOI cannot have best photocatalytic activity. Hence in future, BiOI-based composites [101–104] have to be developed to improve the photocatalytic efficiency by controlling the electron hole recombination and such goal can be achieved only through development of nanocomposite catalyst by adding metals likes tungsten, titanium, molybdenum, iron, zink, to Bi_2O_3 [99, 105–108]. Such catalysts like Bi_2WO_6, $Bi_4Ti_3O_{12}$, $Bi_{12}TiO_{20}$, Bi_2MoO_6, $BiFeO_3$, $BiPO_4$ and Bi_3NbO_7 have gained a lot interest recently by many research groups around the world for removal of heavy metals (Cr), organic pollutants like dyes (RhB) due to enhanced photocatalytic performance. This trend will continue in the future as well.

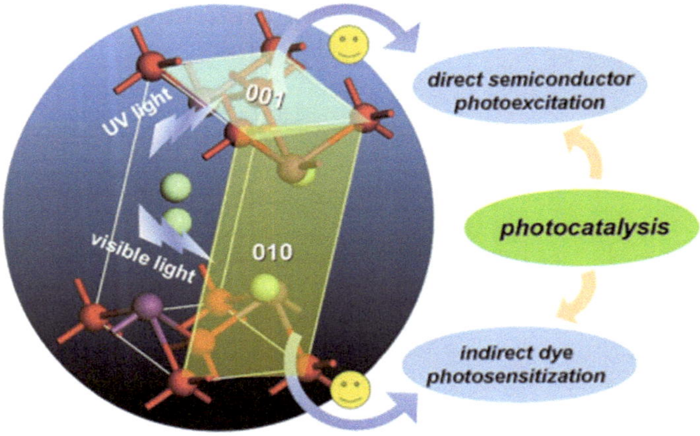

Fig. 8.2 Schematic diagram of facet-dependent photoreactivity of BiOCl single-crystalline nanosheets. Reprinted with permission from Ref. [100]

8.4 Heterojunction Structures of Bismuth Oxyhalides

The heterojunction structures are also very effective method to increase the photo-catalytic action of BiOX. The building of semiconductor hetero-junction allows the incorporation of multiple functional mechanisms along with the contribution of different components, which eventually results in higher photo-catalytic activity/efficiency as compared with the solitary semiconductor [99, 105–130]. The nanostructured BiOX could play three roles in the semiconductor heterojunction systems. At first, the band redox potentials of BiOX are in very broad; so can easily be equal to the energy level of the semiconductor to build an interfacial electric field beneficial for the separation and transportation of photo generated electron–hole pairs. Some of these active semiconductors are AgI/BiOI, Bi_2WO_6/BiOI, C_3N_4/BiOBr, and AgBr/BiOBr [106–119]. At second, the narrow bandgap BiOX semiconductors such as BiOI and BiOBr could be applied to photosensitize other wide bandgap semiconductors for the efficient use of sun light i.e. we could couple BiOI with ZnO and TiO_2 to build p–n heterostructures at low temperatures through facile chemical bath method. TiO_2/BiOI and ZnO/BiOI heterostructures's morphology, chemical composition, and light response range can be adjusted by controlling the molar ratios of Bi to Ti/Zn in the precursor's agents. A more negative conduction band value of BiOI as compared with TiO_2 and ZnO could persuade the effective transfer of photo-generated electrons from BiOI to TiO_2/ZnO, which extends the absorption wavelength range of TiO_2 and ZnO, and also effectively prevents electron–hole pairs recombination. Thirdly, the structural diversity of BiOX permits them to be attached with carbon-based nanomaterials, particularly graphene, to enhance the charge transfer and separation [109, 110]. The hetero structures linkage of graphene with BiOX, including BiOCl/graphene, BiOBr/graphene, and

BiOI/graphene, have been prepared by the use of strong chemical bonding between BiOX and graphene [111–128]. Most of the time, graphene functioned as a co-catalyst to extract photo generated electrons. Therefore, facile electron transfer from BiOX to graphene can significantly prevent the electron–hole recombination and thus increase the lifetime of charge carriers, finally augmenting the photocatalytic efficiency.

In summary, the future work will be focused more on the morphology and structure mediated Bismuth-related nanomaterials, low dimensional shape dependent bismuth-related nanostructures, facet-controlled Bismith related semiconductors nanomaterials, hierarchical bismuth related nanostructures, composited bismuth-related nanomaterials, composited with noble metals, element doped bismuth related nanomaterials specially doped with metals and non metals for photo-catalytic applications.

In spite of the fact that a great deal of work is under advancement on layered structures of BiOX some major issues in arrangement of the layered-structures and properties of BiOX still should be determined. The real hypothetical models and estimations are still under procedure to illuminate the elements of an inside electric field during photo-catalysis and to clarify the extensive exchange pathway of photoinduced electrons and holes inside the layered structure under the impact of Internal electric field (IEF). Notwithstanding the precious crystal facet engineering, more productive procedures to tune the IEF force ought to be investigated, which will help the improvement of BiOX nanomaterials further. (2) The connections between's the different nanostructures of BiOX and their photocatalytic action are still hazy, albeit escalated endeavors have been given to the synthesis and design of very much characterized nanostructures to advance the photoreactivity of BiOX. Further precise examinations are expected to pick up top to bottom understanding into the structure–property relationship of BiOX nanomaterials. (3) Furthermore the {001} and {010} feature leading bismuth oxyhalides, the organized synthesis of BiOX uncovered with other high-energy crystal facets are very attractive in light of the fact that nuclear strides, dangling bonds, crimps and edges on responsive high-vitality face can go about as dynamic locales to encourage the adsorption of reactant particles, surface exchange between photoexcited electrons and reactant atoms, and the desorption of item particles, hence supporting the photocatalytic response. Additionally, the fruitful synthesis of bismuth oxyhalides uncovered with various crystal facets features permits us to further study their aspect subordinate properties in more detail. (4) The redox way of various crystal facets of bismuth oxyhalide nanomaterials still stays hazy. As the movement of photograph initiated electrons and openings are associated nearly to the aspect presentation, there is a dire interest to clear up the feature related redox conduct of bismuth oxyhalide nanomaterials, which can give compelling direction to the outline and combination of superior photocatalysts with particular aspect introduction for critical oxidative or reductive responses. (5) The current photocatalytic utilizations of bismuth oxyhalide nanomaterials basically concentrate on natural applications. It will be important to extend their photocatalytic applications to water splitting, CO_2 photo-reduction, and particular oxidation of natural mixes. Contrasted with conventional photo-catalysts, for example, TiO_2 and ZnO, bismuth oxyhalide photo-catalysts are still in their initial stages, which has a considerable measure of space for further advancement. We trust this book on the BiOX combination, layered-structure intervened properties and photoactivity upgrade methodologies of layered bismuth oxyhalide nano-photocatalysts will have unquestionably some effect on the advancement of BiOX and photocatalytic utilizations of BiOX nanomaterials in the fields of environmental remediation, water splitting and energy harvesting.

Keeping in view of significance of BiOX catalysts and their vital applications in the field of waste water treatment, hydrogen generation by water splitting, reduction of CO_2 into high value hydrocarbons and solar energy harvesting, there is a strong interest in future for the other family of catalysts using Bismith compounds. Here the future trends seems Bimuth Vanandates ($BiVO_4$). Such catalysts can be found in three crystalline forms like tetragonal zircon, monoclinic scheelite and tetragonal scheelite structures. From them, Monoclinic $BiVO_4$ (m-$BiVO_4$) having narrow band gap of 2.4 eV had strong visible absorption along with UV band. Hence its visible light absorption property makes this catalyst very significant.

References

1. Gondal MA, Adeseda AA, Rashid SG, Hameed A, Aslam M, Ismail IMI, Baig U, Dastageer MA, Al-Arfaj AR, Rehman A (2016) Facile synthesis, characterization and photocatalytic performance of Au–Ag alloy nanoparticles dispersed on graphitic carbon nitride under visible light irradiations. J Mol Catal A 423:114–125
2. Seddigi ZS, Gondal MA, Rashid SG, Abdulaziz MA, Ahmed SA (2016) Facile synthesis and catalytic performance of nanosheet—nanorods g-C_3N_4-Bi_2WO_6 heterojunction catalyst and effect of silver nanoparticles loading on bare Bi_2WO_6 and g-C_3N_4-Bi_2WO_6 for N-deethylation process. J Mol Catal A 420:167–177
3. Aslam M, Tahir Soomro M, Ismail Iqbal M I, Qari Huda A, Gondal MA, Hameed A (2015) The facile synthesis, characterization and evaluation of photocatalytic activity of bimetallic $FeBiO_3$ in natural sunlight exposure. RSC Adv 2015(5):102663
4. Ahmad N, Gondal MA, Sheikh AK (2016) Comparative study of different solar based photo catalytic reactors for disinfection of contaminated water. Desalin Water Treat 57(1):213–220
5. Seddigi ZS, Ahmed SA, Bumajdad A, Gondal MA, Danish EY, Shawky AM, Yarkandi NH (2016) Photocatalytic degradation of tert-butyl alcohol and tert-butyl formate using palladium-doped zinc oxide nanoparticles with UV-irradiation. Desalin Water Treat 57/2
6. Gondal MA, Dastageer MA, Khalil A, Rashid SG, Baig U (2015) Photo-catalytic deactivation of sulfate reducing bacteria—a comparative study with different catalysts and the preeminence of Pd-loaded WO_3 nanoparticles. RSC Adv 2015(5):51399–51406
7. Aslam M, Qamar MT, Tahir Soomro M, Ismail IMI, Salah N, Almeelbi T, Gondal MA, Hameed A (2016) The effect of sunlight induced surface defects on the photocatalytic activity of nanosized CeO_2 for the degradation of phenol and its derivatives. Appl Catal B 180:391–402
8. Baig U, Gondal MA, Alam MF, Laskar AA, Alam M, Younus H (2015) Enzyme immobilization and molecular modeling studies on organic-inorganic polypyrrole titanium (IV) phosphate nanocomposite. New J Chem 2015(39):6976–6986

9. Owolabi TO, Gondal MA (2015) Estimation of surface tension of methyl esters biodiesels using computational intelligence technique. Appl Soft Comput J 37:227–233
10. Aslam M, Tahir Soomro M, Ismail IMI, Salah N, Gondal MA, Hameed A (2015) Sunlight mediated removal of chlorophenols over tungsten supported ZnO: electrochemical and photocatalytic studies. J Environ Chem Eng 3(3):1901–1911
11. Baig U, Rao RAK, Khan AA, Sanagi MM, Gondal MA (2015) Removal of carcinogenic hexavalent chromium from aqueous solutions using newly synthesized and characterized polypyrrole-titanium(IV) phosphate nanocomposite. Chem Eng J 280(2015):494–504
12. Rashid SG, Gondal MA, Hameed A, Aslam M, Dastageer MA, Yamani ZH, Anjum DH (2015) Synthesis, characterization and visible light photocatalytic activity of Cr^{3+} and Ce^{3+} co-doped TiON for the degradation of humic acid. RSC Adv 5(41):32323–32332, 9
13. Randhawa MA, Gondal MA, AL-Zahrani HJ, Rashid SG, Ali MA (2015) Synthesis, morphology and antifungal activity of nano particulated amphotericin-B, ketoconazole and thymoquinone against Candida albicans yeasts and Candida-biofilm. J Environ Sci Health A 50:119–124
14. Chang X, Zheng J, Gondal MA, Ji G (2015) Photocatalytic conversion of CO_2 into value-added hydrocarbon (methanol) with high selectivity over ZnS nanoparticles driven by 355-nm pulsed laser. Res Chem Intermed 41:739–747
15. Gondal MA, Adesida AA, Rashid SG, Shi S, Khan R, Yamani ZH, Shen K, Xu Q, Seddigi ZS, Chang X (2015) Preparation of WO_3/g-C_3N_4 composites and their enhanced photodegradation of Rhodamine B in aqueous solution under visible light irradiation114(1). React Kinet, Mech Catal 114(1):357–367
16. Seddigi ZS, Ahmed SA, Bumajdad A, Danish EY, Gondal MA, Shwaky AM, Soylak M (2015) The efficient photocatalytic degradation of methyl tert-butyl ether under Pd/ZnO and visible light irradiation. Photochem Photobiol 91:265–271
17. Ji G, Liu Y, Zhu L, Wang J, Zhang B, Chang X, Gondal MA, Dastageer MA (2014) High-active direct Z-scheme Si/TiO_2 photocatalyst for boosted CO_2 reduction into value-added methanol. RSC Adv 4(100):56961–56969
18. Aslam HM, Ismail MI, Gondal MA (2014) Sunlight assisted photocatalytic mineralization of nitrophenol isomers over W^{6+} impregnated ZnO. Appl Catal B, Environ 160–161(1):227–239
19. Gondal MA, Fajgar R, Chang X, Shenb K, Xu Q (2014) ArF excimer laser-induced deposition of Ag/C nanocomposite thin films in the presence of n-Hexane. Appl Surf Sci 311 (2014):95–100
20. Shen K, Gondal MA, Chang X, Ali MA (2014) Batch and column adsorption of dye contaminants using a low-cost sand adsorbent. Res Chem Intermed 113(Sept issue)
21. Wang J, Ji G, Liu Y, Gondal MA, Chang X (2014) Cu_2O/TiO_2 hetero structure nanotube arrays prepared by an electro-deposition method exhibiting enhanced photo-catalytic activity for CO_2 reduction to methanol. Catal Commun 46:17–21
22. Shen K, Gondal MA, Rashid RG, Shi S, Wang S, Sun J, Su Q (2014) Preparation of ternary Ag/Ag_3PO_4/g-C_3N_4 hybrid photocatalysts and their enhanced photocatalytic activity driven by visible light. Chin J Catal 35:78–84
23. Gondal MA, Siddiqui MN (2014), Nano-catalysts supported laser-induced photocatalytic degradation of MTBE. J Environ Sci Health A 49(1):52–58
24. Hameed A, Aslam M, Ismail MI, Gondal MA (2014) Photocatalytic conversion of methane into methanol: performance of silver impregnated WO_3. Appl Catal A 470(2014):327–335
25. Ali Z, Aslam M, Ismail IM, Hameed A, Hussain ST, Chaudhry MN, Gondal MA (2014) Syntheis, characterization and photocatalytic activity AL_2O_3-TiO_2-based composites. J Environ Sci Health A 49(1):125–134
26. Gondal MA, Ali MA, Dastageer MA, Chang C (2013) CO_2 conversion into methanol using granular silicon carbide (α6H–SiC): a comparative evaluation of 355 nm laser and xenon mercury broad band radiation sources. Catal Lett 193:108–117
27. Li L, Gondal MA, Sun J, Shen K, Chang X (2013) Significant enhancement in removal of Methyl Orange from aqueous solution in the presence of Rhodamine B using Ag_3PO_4 photo-catalyst. Energy Environ Focus 2:188–194

28. Gondal MA, Rashid SG, Dastageer MA, Zubair SM, Ali MA, Lienhard JH, McKinley GH, Varanasi KK (2013) Sol-gel synthesis of Au/Cu–TiO_2 nanocomposite and their morphological and optical properties. IEEE J Photonics 5(3):2013
29. Gondal MA, Dastageer MA, Rashid SG, Zubair SM, Ali MA, Anjum DH, Lienhard JH, McKinley GH, Varanassi K (2013) Plasmonic resonance enhanced photo-catalysis in the visible region with Au/Cu:TiO_2 nanocomposites—prospect for solar energy utilization for various applications: case study of Cr(VI) removal from water. Sci Adv Mater 5:1–8
30. Gondal MA, Chang X, Sha WEI, Yamani ZH, Zhou Q (2013) Enhanced photoactivity on Ag/Ag_3PO_4 composites by plasmonic effect. J Colloid Interface Sci 392:325–330
31. M. A. Gondal, M.A. Ali, M.A.Dastageer, X.Chang (2012), CO_2 conversion into methanol using granular silicon carbide (α6H-SiC): A comparative evaluation of 355 nm Laser and Xenon mercury broad band radiation sources, Catalysis Letters Volume 143, pp 108–117, 2013
32. Gondal MA, Li C, Chang X, Sikong L, Yamani ZH, Zhou Q, Yang F, Lin Q (2012) Facile preparation of magnetic C/TiO_2/Ni composites and their photocatalytic performance for removal of a dye from water under UV light irradiation. J Environ Sci Health A 47(570–576):2012
33. Gondal MA, Ali MA, Chang XF, Shen K, Xu QY, Yamani ZH (2012) Pulsed laser induced photocatalytic reduction of greenhouse gas CO_2 into methanol: a value-added hydrocarbon product over SiC. J Environ Sci Health A 47(11):1571–1576
34. Gondal MA, Randhawa MA, Alzahrani AJ, Siddqui MN (2012) Morphology and antifungal effect of nano ZnO and nano Pd doped nano ZnO against Aspergillus and Candida. J Environ Sci Health 47(10):1413–1418
35. Liang X, Gondal MA, Chang X, Yamani ZH, Li N, Lu H, Ji G (2011) Facile preparation of magnetic separable powdered-activated-carbon/Ni adsorbent and its application in removal of perfluorooctane sulfonate (PFOS) from aqueous solution. J Environ Sci Health Part A 46:1482–1490
36. Qamar M, Yamani ZH, Gondal MA, Alhoshiani K (2011) Synthesis and comparative photocatalytic activity of Pt/WO_3 and Au/WO_3 nanocomposites under sunlight-type excitation. Solid State Sci 13(9):1748–1754
37. Qamar M, Gondal MA, Yamani ZH (2011) Synthesis of nanostructured NiO and its application in laser-induced photocatalytic reduction of Cr(VI) from water. J Mol Catal A 343(83–88):2011
38. Siddiqui MN, Gondal MA (2011) Laser based photo-oxidative degradation of methyl tertiary-butyl ether (MTBE) using zinc oxide (ZnO) catalyst. J Environ Sci Health Part A 46:1154–1159
39. Khalil A, Gondal MA, Dastageer MA (2011) Augmented photo catalytic activity Of palladium doped nano ZnO in the disinfection of *Escherichia coli* microorganism from water. Appl Catal A 402(1–2):162–167
40. Bagabas A, Gondal M, Khalil A, Dastageer A, Yamani Z, Ashameri M (2010) Laser-induced photocatalytic inactivation of coliform bacteria from water using pd-loaded nano-WO_3. Stud Surf Sci Catal 175:279–282
41. Saleh TA, Gondal MA, Drmosh QA, Yamani ZH, Yamani A (2011) Enhancement in photocatalytic activity for acetaldehyde removal by embedding ZnO nano particles on multiwall carbon nanotubes. Chem Eng J Chem Eng J 166:407–412
42. Qamar M, Gondal MA, Yamani ZH (2011) Laser-induced efficient removal of Cr (VI) catalyzed by ZnO semiconductor. J Hazard Mater 187(2011):258–263
43. Hayat K, Gondal MA, Khaled MM, Ahmed S, Shemsi AM (2011) Nano ZnO synthesis by modified Sol Gel method and its application in heterogeneous photo-catalysis removal of phenol from water. Appl Catal A 393:122–129
44. Gondal MA, Chang X, Yamani ZH, Yang G, Ji G (2011) Photocatalytic removal of Sulforhodamine B from aqueous solution on GaN thin films under UV pulsed laser irradiation. J Environ Sci Health Part A 46(4):415–419

45. Gondal MA, Khalil A, Dastageer MA, Yamani ZH (2011) Parametric optimization of nano-ZnO photo catalyst for effective disinfection of *Escherichia coli* micro organism in Water. J Nanopart Res 13(8):3423–3430
46. Hayat K, Gondal MA, Khaled MM, Yamani ZH, Ahmed S (2011) Laser induced photocatalytic degradation of hazardous dye (Safranin-O) using self synthesized nanocrystalline WO_3. J Hazard Mater 186(2–3):1226–1233
47. Saleh TA, Gondal MA, Drmosh QA (2010) Laser induced photocatalytic activity of locally synthesized CNT/ZnO nanocomposite for removal of cyanide from water. Nanotechnology 21(49):10. Article number 495705
48. Gondal MA, Chang X, Yamani ZH (2010) Pulsed laser -induced photocatalytic removal of Rhodamine 6G over BiOCl from aqueous solution. Chem Eng J 165:250–257
49. Hayat K, Gondal MA, Khaled Mazen M (2010) Effect of operational key parameters on photocatalytic degradation of phenol using nano nickel oxide synthesized by sol gel method. J Mole Catal A 336(2011):64–71
50. Hayat K, Gondal MA, Khaled MM, Ahmed S (2010) Kinetic study of laser induced photocatalytic degradation of dye (alizarin yellow) from waste water using nanostructured ZnO. J Environ Sci and Health Part A 45:1413–1420
51. Fazal A, Al-Fayez S, Abdel-Rahman LH, Seddigi ZS, Al-Arfaj AR, El Ali B, Dastageer Mohammad A, Gondal MA (2010) Mixed-ligand complexes of copper(I) with diimines and phosphines: Effective catalysts for the coupling of phenylacetylene with halobenzene. Polyhedron 28:4072–4076
52. Qamar M, Gondal MA, Yamani ZH (2010) Removal of Rhodamine 6G induced by laser and catalyzed by Pt/WO_3 nanocomposite. Catal Commun 11(2010):768–772
53. Randhawa MA, Al-Zahrani AJ, Gondal MA, Bagabas AA (2010) Synthesis, characterization and antimicrobial activity of nano ZnO and Pd loaded nano Zno against enteric pathogens. J Mater Sci Eng 4(5)
54. Gondal MA, Hayat K, Khalid MM, Ahmed S (2010) Photocatalytic removal of hazardous dye from water using nanostructured WO_3. Int J Nano Part 4(1):53–63
55. Gondal MA, Bagabas AA, Dastageer MA, Khalil A (2010) Synthesis, characterization, and antimicrobial application of nano-palladium doped nano-WO_3. J Mole Catal A 323:78–83
56. Qamar M, Gondal MA, Yamani ZH (2009) Synthesis of highly active nanocrystalline WO_3 and its application in laser-induced photocatalytic removal of a dye from water. Catal Commun 10:1980–1984
57. Qamar M, Gondal MA, Hayat K, Yamani ZH, Al-Hooshani K (2009) Laser-induced removal of dye derivative eosin yellowish catalyzed by n-type WO_3 semiconductor catalyst. J Hazard Mater, V 170:584–589
58. Gondal MA, Sayeed MN, Yamani ZH, Arfaj A (2009) Efficient removal of phenol from water using Fe_2O_3 semiconductor catalyst under UV laser irradiation. J Environ Sci Health Part A A44(5):515–521
59. Gondal MA, Sayeed MN (2008) Laser enhanced photocatalytic degradation of organic pollutants from water using ZnO semiconductor catalyst. J Environ Sci Health Part A 43(1):70–77
60. Gondal MA, Sayeed MN, Seddighi Z (2008) Laser enhanced photo-catalytic removal of phenol from water using P-type NiO semiconductor catalyst. J Hazard Mater 155:83–89
61. Gondal MA, Khalil A (2008) Rapid disinfection Of E-Coliforms contaminated water using WO_3 semiconductor catalyst by laser induced photo-catalytic process. J Environ Sci Health Part A 43(5):488–494
62. Gondal MA, Sayeed MN, Arfaj A (2007) Activity comparison of Fe_2O_3, NiO, WO_3 and TiO_2 semiconductor catalysts in phenol degradation by laser enhanced photo-catalytic process. Chem Phys Lett 445:325–330
63. Gondal MA, Seddighi Z (2005) Laser induced photo-catalytic removal of phenol using n-type WO_3 semiconductor catalyst. Chem Phys Lett 417:124
64. Hameed A, Gondal MA, Yamani ZH, Yahya AH (2005) Significance of pH Measurements in photocatalytic splitting of water by 355 nm UV laser. J Mol Catal A, V 227:241

65. Hameed A, Gondal MA (2005) Production of hydrogen-rich syngas using p-type NiO catalyst: a laser-based photocatalytic approach. J Mol Catal A 233:35
66. Gondal MA, Hameed A, Yamani ZH (2005) Laser photocatalytic splitting of water over WO3 catalyst. Energy Sources 27:1151–1165
67. Gondal MA, Hameed A, Yamani ZH, Al-Suwaiyan A (2004) Production of hydrogen and oxygen by water splitting using laser photo-catalysis, over Fe_2O_3. Appl Catal A 268:159
68. Gondal MA, Hameed A, Yamani ZH, Al-Suwaiyan A (2004) Laser induced photocatalytic oxidation of water: activity comparison of α-Fe_2O_3, WO_3, TiO_2 and NiO catalysts. Chem Phys Lett 385:111
69. Yahaya H, Gondal MA, Hameed A (2004) Selective laser enhanced photocatalytic reduction of CO_2 into methanol. Chem Phys Letts V 400:206
70. Hameed A, Gondal MA (2004) Laser induced photocatalytic generation of hydrogen and oxygen Over NiO and TiO_2. J Mol Catal 219:109
71. Hameed A, Gondal MA, Yamani ZH (2004) Effect of transition metal doping on photocatalytic activity of WO_3 under laser illumination: role of 3d-orbitals. Catal Commun 5:715
72. Gondal MA, Hameed A, Yamani ZH (2004) Hydrogen generation by laser transformation of methanol using n-type WO_3 semiconductor catalyst. J Mol Catal A 222:259
73. Gondal MA, Hameed A, Yamani ZH, Arfaj A (2004) Photocatalytic transformation of methane into methanol under UV laser irradiation over WO_3, TiO_2 and NiO catalysts. Chem Phys Lett 392:377
74. Gondal MA, Hameed A, Al-Suwaiyan A (2003) Photocatalytic conversion of methane into methanol using visible laser. Appl Catal 243:165–174
75. Fujishima KH (1972) Electrochemical photolysis of water at a semiconductor electrode. Nature 238:37–38
76. Yang D, Gondal MA, Yamani ZH, Xu Q, Xiang D, Mao J, Shen K (2016) Enhanced photosensitization decomposition of Rhodamine B onto BiOCl nanosheets with controllable-exposed {001} facets. Nanosci Nanotechnol Lett 8:1–8
77. Chang X, Wang S, Qi Q, Gondal MA, Rashid SG, Gao S, Yang D, Shen K, Xu Q, Wang P (2015) Insights into the growth of bismuth nanoparticles on 2D structured BiOCl photocatlyst: an insitue TEM investigation. Dalton Trans 44:15888
78. Chang X, Wang S, Qi Q, Gondal MA, Rashid SG, Yang D, Dastageer MA, Shen K, Xue Q, Wang P (2015) Constrained growth of ultrasmall BiOCl nanodiscs and their enhanced photoreactivity under visible light irradiation. Appl Catal B 10/2015; 176–177:201–211
79. Shi H, Gondal MA, Al-Saadi AA, Chang X (2015), Visible-light-induced photo-degradation enhancement of methyl orange over bismuth oxybromide through a semiconductor mediated process. J Adv Oxid Technol 18(1):78–84
80. Shi S, Gondal MA, Rashid SG, Qi Q, Al-Saadi AA, Yamani ZH, Sui Y, Xu Q, Shen K (2014) Synthesis of g-C_3N_4/$BiOCl_xBr_{1-x}$ hybrid photocatalysts and the photoactivity enhancement driven by visible light. Colloids Surf A 461:202–211
81. Shi S, Gondal MA, Al-Saadi AA, Fajgar R, Kupcik J, Chang X, Shen K, Xu Q, Seddigi ZS (2014) Facile preparation of g-C_3N_4 modified BiOCl hybrid photocatalyst and vital role of frontier orbital energy levels in photo-activity enhancement for removal of model compounds. Colloid Interface Sci 416:212–219
82. Kai Shen MA, Gondal AA, Al-Saadi Li Liye, Chang X, Xu Q (2015) Visible light induced photodegaradtion of Rhodamine dyes over BiOCl and the vital importance of frontier orbital energy of the dye molecules in the reaction kinetics. Res Chem Intermed 41:2753–2766
83. Zhang B, Ji G, Gondal MA, Liu Y, Zhang X, Chang X, Li N (2013) Rapid adsorption properties of flower-like BiOI nanoplates synthesized via a simple EG-assisted solvothermal process. J Nanopart Res 15:1773
84. Shen K, Gondal MA, Xu Q, Shi S, Chang X (2013) Synthesis of RhB/BiOBr hybrid photocatalyst and its utilization in enhanced degradation of methyl orange via visible-light induced photosensitization process. J Adv Oxid Technol 17(1):121–126

85. Zhang B, Ji G, Liu Y, Gondal MA, Chang X (2013) Efficient adsorption and photocatalytic performance of flower-like three-dimensional (3D) Idoped BiOClBr photocatalyst. Catal Commun 36:25–30
86. Shen K, Gondal MA, Li Z, Li L, Xu Q, Yamani ZH (2013) 450 nm visible light-induced photosensitized degradation of Rhodamine B molecules over BiOBr compound from aqueous solution. React Kinet, Mech Catal 109:247–258
87. Luo L, Shen K, Xu Q, Qin Zhou W, Wei MA Gondal (2013) Preparation of multiferroic Co substituted $BiFeO_3$ with enhanced coercive force and its application in sorption removal of dye molecules from aqueous solution. J Alloys Compd 558:73–76
88. Zhang J, Gondal MA, Wei W, Zhang T, Xu Q, Shen K (2012) Preparation of room temperature ferromagnetic $BiFeO_3$ and its application as an highly efficient magnetic separable adsorbent for removal of Rhodamine B from aqueous solution. J Alloys Compd 530:107–110
89. Zhang J, Chang X, Gondal MA, Wei W, Zhang T, Xu Q, Shen K (2012) Synthesis and photocatalytic activity of graphene/BiOBr composites under visible light. Appl Surf Sci 258 (20):7795–8322
90. Gondal MA, Chang X, Al-Saadi AA, Yamani ZH, Zhang J, Ji G (2012) BiOCl-assisted photodegradation of Rhodamine B under white light and monochromatic green pulsed laser irradiation. J Environ Sci Health A 47(8):1192–1200
91. Chang X, Gondal MA, Al-Saadi AA, Shen H, Ali MA, Zhou Q, Zhang J, Du M, Liu 1 Y, Ji G (2012) Photo-degradation of Rhodamine B over unexcited semiconductor compounds of BiOCl and BiOBr. J Colloid Interface Sci 377:290–298
92. Gondal MA, Chang X, Ali MA, Yamani ZH, Ji G (2011) Adsorption and degradation performance of Rhodamine B over BiOBr under 532 nm pulsed laser exposure. Appl Catal A 397:192–200
93. Li TB, Chen G, Zhou C, Shen ZY, Jin RC, Sun JX (2011) New photocatalyst BiOCl/BiOI composites with highly enhanced visible light photocatalytic performances. Dalton Trans 40:6751
94. Liu YY, Son WJ, Lu JB, Huang BB, Dai Y, Whangbo MH (2011) Composition dependence of the photocatalytic activities of $BiOCl_{1-x}Br_x$ solid solutions under visible light. Chem-Eur J 17:9342
95. Jia ZF, Wang FM, Xin F, Zhang BQ (2011) Simple solvothermal routes to synthesize 3D $BiOBr_xI_{1-x}$ microspheres and their visible-light-induced photocatalytic properties. Ind Eng Chem Res 50:6688
96. Cao J, Xu BY, Luo BD, Lin HL, Chen SF (2011) Efficient photocatalytic degradation of bisphenol A and dye pollutants over $BiOI/Zn_2SnO_4$ heterojunction photocatalyst. Catal Commun 13:63
97. Zhuo Y, Huang J, Cao L, Ouyang H, Wu J (2013) Photocatalytic activity of snow-like Bi_2WO_6 microcrystalline for decomposition of Rhodamine B under natural sunlight irradiation. Mater Lett 90:107
98. Cui Z, Liu J, Zeng D, Liu H, Xie C (2010) Quasi-one-dimensional bismuth tungsten oxide nanostructures templated by cotton fibers. J Am Ceram Soc 93:1479
99. Huang J, Zhang H, Zhou X, Zhong X (2013) Dimensionality-dependent performance of nanostructured bismuth sulfide in photodegradation of organic dyes. Mater Chem Phys 138:755
100. Jiang J, Zhao K, Xiao XY, Zhang LZ (2012) Synthesis and facet-dependent photoreactivity of BiOCl single-crystalline nanosheets. J Am Chem Soc 134:4473
101. Shang M, Wang W, Sun S, Zhou L, Zhang L (2008) Bi_2WO_6 nanocrystals with high photocatalytic activities under visible light. J Phys Chem C 112:10407
102. Wu L, Bi J, Li Z, Wang X, Fu X (2008) Rapid preparation of Bi_2WO_6 photocatalyst with nanosheet morphology via microwave-assisted solvothermal synthesis. Catal Today 131:15
103. Zhang L, Bahnemann D (2013) Synthesis of nanovoid Bi_2WO_6 2D ordered arrays as photoanodes for photoelectrochemical water splitting. Chemsuschem 6:283

104. Xiong J, Cheng G, Li G, Qin F, Chen R (2011) Well-crystallized square-like 2D BiOCl nanoplates: mannitol-assisted hydrothermal synthesis and improved visible-light-driven photocatalytic performance†. RSC Adv 1:1542
105. Li G, Ding Y, Zhang Y, Lu Z, Sun H, Chen R (2011) Microwave synthesis of $BiPO_4$ nanostructures and their morphology-dependent photocatalytic performances. J Colloid Interface Sci 363:497
106. Fu J, Tian YL, Chang BB, Xi FN, Dong XP (2012) BiOBr–carbon nitride heterojunctions: synthesis, enhanced activity and photocatalytic mechanism. J Mater Chem 22:21159
107. Zhang X, Zhang LZ, Xie TF, Wang DJ (2009) Low-temperature synthesis and high visible-light-induced photocatalytic activity of BiOI/TiO_2 heterostructures. J Phys Chem C 113:7371
108. Jiang J, Zhang X, Sun PB, Zhang LZ (2011) ZnO/BiOI heterostructures: photoinduced charge-transfer property and enhanced visible-light photocatalytic activity. J Phys Chem C 115:20555
109. Cheng HF, Huang BB, Dai Y, Qin XY, Zhang XY (2010) One-step synthesis of the nanostructured AgI/BiOI composites with highly enhanced visible-light photocatalytic performances. Langmuir 26:6618
110. Li HQ, Cui YM, Hong WS (2013) High photocatalytic performance of BiOI/Bi_2WO_6 toward toluene and Reactive Brilliant Red. Appl Surf Sci 264:581
111. Kong L, Jiang Z, Lai HH, Nicholls RJ, Xiao TC, Jones MO, Edwards PP (2012) Unusual reactivity of visible-light-responsive AgBr–BiOBr heterojunction photocatalysts. J Catal 293:116
112. Xiang QJ, Yu JG, Jaroniec M (2012) Graphene-based semiconductor photocatalysts. Chem Soc Rev 41:782
113. Tu WG, Zhou Y, Zou ZG (2013) Versatile graphene-promoting photocatalytic performance of semiconductors: basic principles, synthesis, solar energy conversion, and environmental applications. Adv Funct Mater 23:4996
114. Gao FD, Zeng DW, Huang QW, Tian SQ, Xie CS (2012) Chemically bonded graphene/BiOCl nanocomposites as high-performance photocatalysts. Phys Chem Chem Phys 14:10572
115. Song SY, Gao W, Wang X, Li XY, Liu DP, Xing Y, Zhang HJ (2012) Microwave-assisted synthesis of BiOBr/graphene nanocomposites and their enhanced photocatalytic activity. Dalton Trans 41:10472
116. Tu XM, Luo SL, Chen GX, Li JH (2012) One-pot synthesis, characterization, and enhanced photocatalytic activity of a BiOBr–graphene composite. Chem-Eur J 18:14359
117. Liu H, Su Y, Chen Z, Jin ZT, Wang Y (2014) Graphene sheets grafted three-dimensional $BiOBr_{0.2}I_{0.8}$ microspheres with excellent photocatalytic activity under visible light. J Hazard Mater 266:75
118. Liu H, Cao WR, Su Y, Chen Z, Wang Y (2013) Bismuth oxyiodide-graphene nanocomposites with high visible light photocatalytic activity. J Colloid Interface Sci 398 (161):204
119. Liu Z, Xu WC, Fang JZ, Xu XX, Wu SX, Zhu XM, Chen ZH (2012) Decoration of BiOI quantum size nanoparticles with reduced graphene oxide in enhanced visible-light-driven photocatalytic studies. Appl Surf Sci 259:441
120. Ai ZH, Ho W, Lee SC (2011) Efficient visible light photocatalytic removal of NO with BiOBr-graphene nanocomposites. J Phys Chem C 115:25330
121. Zhang XM, Chang XF, Gondal MA, Zhang B, Liu YS, Ji GB (2012) Synthesis and photocatalytic activity of graphene/BiOBr composites under visible light. Appl Surf Sci 258:7826
122. Ozawa H, Sakai K (2011) Photo-hydrogen-evolving molecular devices driving visible-light-induced water reduction into molecular hydrogen: structure–activity relationship and reaction mechanism. Chem Commun 47:2227
123. Osterloh FE (2013) Inorganic nanostructures for photoelectrochemical and photocatalytic water splitting. Chem Soc Rev 42:2294

124. Dong F, Sun YJ, Fu M, Wu ZB, Lee SC (2012) Room temperature synthesis and highly enhanced visible light photocatalytic activity of porous BiOI/BiOCl composites nanoplates microflowers. J Hazard Mater 219:26
125. Xiao X, Hao R, Liang M, Zuo XX, Nan JM, Li LS, Zhang WD (2012) One-pot solvothermal synthesis of three-dimensional (3D) BiOI/BiOCl composites with enhanced visible-light photocatalytic activities for the degradation of bisphenol-A. J Hazard Mater 233:122
126. Shenawi-Khalil S, Uvarov V, Fronton S, Popov I, Sasson Y (2012) A novel class of heterojunction photocatalysts with highly enhanced visible light photocatalytic performances: $y\mathrm{BiO(Cl}_x\mathrm{Br}_{1-x})$–$(1 - y)$ bismuth oxide hydrate. Appl Catal B 117:148
127. Kim WJ, Pradhan D, Min B, Sohn Y (2014) Adsorption/photocatalytic activity and fundamental natures of BiOCl and $\mathrm{BiOCl}_x\mathrm{I}_{1-x}$ prepared in water and ethylene glycol environments, and Ag and Au-doping effects. Appl Catal B 147:711
128. Zhang B, Ji GB, Liu YS, Gondal MA, Chang XF (2013) Efficient adsorption and photocatalytic performance of flower-like three-dimensional (3D) I-doped BiOClBr photocatalyst. Catal Commun 36:25
129. Shang J, Hao WC, Lv XJ, Wang TM, Wang XL, Du Y, Dou SX, Xie TF, Wang DJ, Wang J (2014) Bismuth oxybromide with reasonable photocatalytic reduction activity under visible light. ACS Catal 4:954
130. Chen HL, Lee WW, Chung WH, Lin HP, Chen YJ, Jiang YR, Lin WY, Chen CC (2014) Controlled hydrothermal synthesis of bismuth oxybromides and their photocatalytic properties. J. Taiwan Inst Chem Eng 45(4):1892–1909. doi:10.1016/j.jtice.2013.12.015